Lecture Notes on
Coastal and Estuarine
Studies

Managing Editors:
Malcolm J. Bowman   Richard T. Barber
Christopher N.K. Mooers   John A. Raven

21

Thomas Stocker
Kolumban Hutter

Topographic Waves
in Channels and Lakes
on the f-Plane

Springer-Verlag
Berlin Heidelberg New York London Paris Tokyo

Authors

Thomas Stocker
Kolumban Hutter
Laboratory of Hydraulics, Hydrology and Glaciology, ETH Zürich
Gloriastr. 37–39, 8092 Zürich, Switzerland

ISBN 978-3-540-17623-7   ISBN 978-3-642-50990-2 (eBook)
DOI 10.1007/978-3-642-50990-2

Thomas Stocker dedicates this book to *Christine*

Kolumban Hutter expresses his appreciation, affection and love by dedicating it to his parents *Jakob and Paula Hutter-Gmür*

cal models. These are named Ball-modes due to their first description by Ball (1965). When the wave motion manifests a large number of small gyres, patterns are similar to topographic waves in infinite channels. Finally, the system also sustains bay-trapped modes, the stream function of which decays exponentially as one moves away from the bay. These bound-states exhibit features not unlike those of Taylor reflections of Kelvin waves on the f-plane: Pure reflection in Taylor's problem corresponds to truly bound states of the topographic wave equation; incomplete Taylor reflection corresponds to a bound state with a far field leakage.

Chapter 1 reviews observational facts as far as ocean bays and lakes are concerned. Completeness is no intention as we prefer to state the controversial facts rather than the established agreement.

In *Chapter 2* we present the governing equations of shallow water theory and deduce the topographic wave equation by successive steps of approximation. The analysis of the homogeneous water body and the two-layer model are known, however we also present the analysis for a continuously stratified water body, which is new. Its result is that to first order the long periodic topographic oscillations drive a baroclinic motion but not vice versa.

Chapter 3 collects known solutions and describes their physical properties for circular and elliptical geometries, shelves, trenches and exterior domains of elliptical islands. Despite the variety of physical behaviors exhibited by these solutions they are not able to satisfactorily explain the observed low frequency response of some smaller lakes and appear to be at variance with alternative computational results.

In *Chapter 4* we therefore propose a projection method by which the spatially two-dimensional topographic wave equation is transformed into a hierarchy of spatially one-dimensional substitute problems. The procedure of derivation of this alternative set of equations is the Method of Weighted Residuals, and the intention is very much in the spirit of Chrystal or Defant, namely to deduce a set of channel equations, now for topographic Rossby waves.

In *Chapter 5* we apply this approximate method to infinite channels and thus demonstrate its workability; numerical solutions are compared with exact analytical results, but we also present many new results for the propagation of topographic waves in channels exhibiting strong topography.

Preface

The last one or two decades have witnessed an increased interest in topographic Rossby waves, both from a theoretical computational as well as
an observational point of view. However, even though long periodic processes were observed in lakes and ocean basins with considerable detail,
it appears that interpretation in terms of physical models is not sufficiently conclusive. The reasons for this lack in understanding may be
sought both, in the insufficient spatial resolution or the brevity of the
time series of the available data and the inadequacy of the theoretical
understanding of long periodic oscillating processes in lakes and ocean
bays. Advancement will emerge from intensified studies of both aspects,
but it is equally our believe that the understanding of long periodic
oscillations in lakes is presently likely to profit most from a theore-
tical-computational study of topographic Rossby waves in enclosed basins.

With this tractate we aim to provide the reader with the basic concepts
of wave motion in shallow waters at subinertial frequencies. Our questions throughout this monograph are essentially: How can the solutions
to this topographic wave equation in a prescribed idealized domain be
construced; what are the physical properties of these solutions; are
their features identifiable by observations; how reliable are such interpretations, etc.? We are obviously not the first ones to be concerned
with these questions (and we will review many aspects of it), but by applying a special numerical integration procedure (a pseudo-spectral or
projection method) we claim to have achieved a considerable advancement
in the understanding of the behavior of topographic Rossby waves in infinite and semi-infinite channels and in closed rectangles. In particular, we prove that besides the free travelling waves, the topographic
wave equation admits bound states; these states consist of wave activity
which is essentially restricted to a limited subdomain of the basins.
Another result is the fact that the spectrum of the topographic wave
operator is highly densely populated and that the modes of these eigenfrequencies may be structurally very different from each other: In other
words, there is no apparent order in the mode structure if the frequencies are ordered according to their magnitude. Nevertheless, scrutiny of
the dispersion relation permits identification of essentially three types of topographic wave behavior in enclosed basins. When the current
structures consist of a few global vortices then they can be identified
with the modes which are well-known and extensively studied by analyti-

Chapter 6 presents a detailed analysis of solutions in rectangular domains. It is demonstrated that the smootheness properties of the isobaths play an important role in establishing the type of possible mode structure. If isobaths end at the shore energy approaching this shore from one side of the lake cannot leak to the other side; similarly a kink in the isobaths appears to considerably hinder the transfer of energy beyond it. However, it is also demonstrated that the long ends of lakes tends to favour localized topographic wave response.

In *Chapter 7* these properties are further scrutinized by studying the reflection of topographic waves in semi-infinite channels. Here it becomes apparent that the localized wave features in Chapter 6 correspond to bound states and that large scale and small scale behavior is associated with phase and group velocity of reflected waves being essentially parallel and anti-parallel, respectively.

Finally *Chapter 8* points into directions of further research.

The timely completion of this manuscript profited from the help of many individuals. F. Langenegger and M. Staub typed the preliminary drafts, and C. Bucher and I. Wiederkehr drew the art work. We are very thankful to them. Our special acknowledgements, however, go to F. Langenegger who is responsible for the composition and layout of the camera ready manuscript. Last, but not least we thank Prof. Dr. D. Vischer, Director of the Laboratory of Hydraulics, Hydrology and Glaciology for his support and his interest in this work.

Zürich, December 1986   Kolumban Hutter, Thomas Stocker

Contents

1. Introduction

1.1 Preamble

"During the past one or two decades there has been an increasing number
of temperature and current observations in various lakes and ocean bays
which show pronounced oscillations with a characteristic period of a few
days. The existence of such long periodic oscillations in a rotating circular basin with a parabolic depth profile was pointed out by Poincaré
in 1910, and the first explicit solution for these topographic waves
(also called second class waves, vortex modes, quasi-geostrophic waves)
was given by Lamb (1932). However, it is only very recently (e.g., see
Saylor et al. 1980) that observations *and* an appropriate theory have been
combined to provide a unified account of low frequency motions in lakes"
(after Mysak, 1984).

On reviewing the literature on topographic waves (hereafter referred to
as TW's) and after a few years of own observational and theoretical work,
it became apparent that there is a need for a summary and synthesis of
the many coherent, yet still partly conflicting results. Whereas three to
four years ago we thought that bringing observations and theory together
was merely a question of straightforward application of numerical techniques to the governing equations, we have since then adopted a more cautious attitude. Our current thinking is that the problem of topographic
waves in lakes and small ocean basins has considerably advanced during
the last few years, but the interpretation of longperiodic signals remains essentially still an open question.

These notes were written as a challenge to todays and future workers in
physical limnology in order to provide them with the state of the art as
we see it and, even more so, to encourage them to improve on our own
techniques and to broaden present knowledge.

1.2 Waves in waters

Waves in open waters, such as the ocean, lakes, channels, etc., arise in
a variety of forms and types and have various physical reasons of their
existence. Rather than to try to define the physical meaning of the no-
tion "wave" we quote here a beautiful statement attributed to Einstein,
that expresses the essentials of what is meant by waves more adequately
than we could have done. The statement *) is:

"Irgend ein Klatsch, der, sagen wir, in Washington aufgebracht wird, ge-
langt sehr rasch nach New York, wenn auch nicht eine einzige von den an
der Weitergabe beteiligten Personen tatsächlich von der einen Stadt in
die andere reist. Wir haben es vielmehr gewissermassen mit zwei ganz ver-
schiedenen Bewegungen zu tun, der des Gerüchtes selbst, das von Washington nach New York dringt, und der jener Personen, die das Gerücht ver-
breiten."

Waves are generated by external forces. These forces are complex in
their spatial and temporal structure and thus impose a large spectrum
of the typical physical scales. Their explanation and relation to the
primary cause is the goal of any wave science. There are basically *two
qualities* which govern wave motion in open waters. Firstly, water exhibits certain *physical properties*, and thus gives rise to mechanical, chemical and electromagnetic response mechanisms; secondly the water is
confined to the geometry of the container where it resides. Thus wave
motions also reflect *geometric properties*. For instance, acoustic waves
are due to the mechanical properties, namely the compressibility of water and are hardly influenced or modified by the "state" of the container. What we mean by this is that the dispersion relation of acoustic
waves is to first order not modified when the container is rotating or
has a complicated topography. Another example of waves governed by the
property of water would be electromagnetic waves, light, of which the
dispersion relation remains unaffected by a possible rotation of the
container.

The larger the "scale" of the driving mechanism of the wave is, the
lower will, in general, be its frequency. Whereas a typical acoustic
wave has a frequency of $10^2 \, s^{-1}$, external gravity waves have about $10^{-3} \, s^{-1}$
and internal (topographic) waves in a lake even $10^{-5} \, s^{-1}$. This shows
that waves in waters occupy a broad frequency spectrum which spans over
many log cycles. *Figure 1.1* provides a first survey of these various
waves which are sustained in waters.

Gravity waves are due to the displacement of the mass of water from an
equilibrium position; the restoring force is gravity, but the speed of
propagation depends also on the local depth of the water basin.

---

*) *"Any gossip which, say, is spread in Washington arrives very quickly in New York,
even though not a single person who participates in the spreading of the rumor
really travels from one city to the other. Rather, we are dealt here with two entirely different motions, that of the rumor itself, which propagates from Washington to New York, and that of the people who are spreading the gossip."*

*We have seen the German quotation at an exhibition "TECHNORAMA" in Winterthur,
Switzerland, but were not able to trace the exact source of the quotation.*

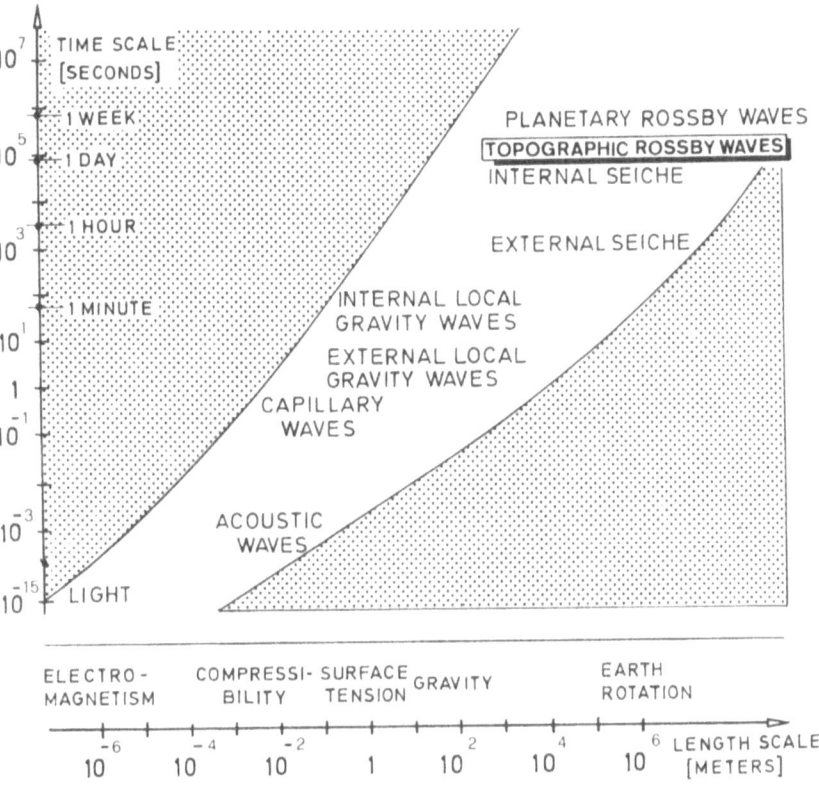

Figure 1.1
Variety of waves occurring in waters with their individual
time and length scales. The wave generating mechanisms are
listed below.

Seiche[*) motions are governed by the shape of the container even to a
larger extent, i.e. the position of the boundaries determines both, frequency and wavelength (in a closed basin) of the seiche. Kelvin and Poincaré waves owe their existence to the rotation of the container and the
Coriolis force which can be felt in such a non-inertial frame. Topographic waves, finally, are governed mainly by the container, in that they
require for their occurrence a container with nontrivial topography and
rotation. Both qualities are natural features of lake basins, in that
they exhibit a variable topography on the rotating earth.

Waves which are mainly meteorologically induced are classified into *gravity waves* and *rotational waves*. Among the former, we distinguish between

---

*) The term "seiche" is used to designate global periodic lake level oscillations. It
appeared for the first time in 1730 in a work by Duillier on the history of Lake
Geneva and is now commonly used in limnology. It is believed to originate from the
french verb "sêcher" (= to dry out).

the long gravity waves and the short surface waves. Short surface waves
represent the most conspicuous oscillatory response of open water to wind
action; they are directly detectable by the eye. Their periods are se-
conds, and their wavelengths are small in comparison to the water depth.
These waves contribute much to the upper turbulent fluctuating motion.
Gravity waves, which affect the global overall response, have often periods of much larger duration; their wavelengths are large as compared
to the water depth and may be comparable to the horizontal dimensions of
the water basin. These waves essentially arise in two different forms,
depending on whether the restoring force is primarily the cause of surface elevations from an undeformed equilibrium level or whether water
particles within a stably stratified fluid are displaced from their
neutral positions.

We call the former *external* or *barotropic*, because the restoring force is
induced at the outer boundary and density variations are irrelevant;
alternatively the latter are called *internal* or *baroclinic*. Typical speeds
of phase propagation in natural lakes are 10-50 ms^{-1} for barotropic gravity waves and 0.1-2 ms^{-1} for baroclinic motions. Periods of oscillations
in water basins of 10 to 100 km extent are therefore minutes or hours
for barotropic processes, but hours and days for baroclinic processes.
These figures remain valid when gravity waves are modified by effects of
the rotation of the Earth, but, of course, the structure of these modified waves (Kelvin-, Poincaré- and Sverdrup- waves) is quite different
from their non-rotational variants.

One property of gravity waves in open waters is that to any such wave
on a rotating frame there exists a counterpart on an inertial frame.
This property characterizes a particular *class* of waves; those belonging
to it are called *first class waves*. Quite differently, *rotational waves* owe
their being to the existence of the rotation of the carrier medium; in
the limit of zero rotation (inertial frame) they cease to exist. Waves
having this property are called *second class waves*. In the context of geophysical fluid mechanics they arise as various forms of the so-called
Rossby waves. Planetary Rossby waves owe their existence to the North-
South variation of the Coriolis parameter, the so-called β-effect. *Topographic Rossby waves* (TW) are based on a constant Coriolis parameter but
require a variable depth; structurally they are more complex than planetary waves, because bathymetric variations can arise in all directions. Finally, there are the combination of the two, yielding plane-
tary-topographic waves. Periods of TW's are long and depend strongly on

the bathymetry of the basins. Typically they are from a few days to weeks, months and - exceptionally - years. Such waves have extensively been studied for open and semi-open domains with variations of the topography restricted to a shore-near band (shelf and trench waves) and with bathymetric profiles that permit analytic treatment of the governing equation (Mysak, 1980).

In this book we concentrate on TW's in confined domains and with depth profiles which may not easily admit explicit analytic solutions.

1.3 Observations - Their interpretations

In this section we describe a series of long-periodic oscillations of temperature or current data indicating that the phenomenon which underlies these observations can probably be interpreted in terms of topographic waves. Our report is not in chronological order, but rather follows easiness in interpretation. The details of the theoretical models (assumptions, approximations etc.) which are used to explain the observations are postponed to chapter 2.

a) Lake Michigan

The analysis outlined here and the figures are due to Saylor, Huang & Reid (1980) and Huang & Saylor (1982). During spring, summer and fall 1976 current meters were deployed in Southern Lake Michigan as indicated in *Figure 1.2a*. Most of these current meters were positioned along the eastern shore between Benton Harbor and Muskegon, but 10 current meters were also deployed along a straight line connecting opposite points along the shores between Racine and Holland.

The bathymetry of the Southern Lake Michigan is simple and concise. An approximation of the topography by a circular basin with radial profile dependence, see *Figure 1.2b*, may not be an oversimplification of the situation at hand.

Figure 1.3 collects kinetic energy spectra of the eastward (*Figure 1.3a*) and northward (*Figure 1.3b*) velocity components (generally at the 25 m level) at the six stations indicated in *Figure 1.2*. Vertical lines are drawn to accentuate the two conspicuous energy peaks at near inertial (~ 17 hours) and near 4-day periods. Thus, the lake responds distinctly at these periods. This is further substantiated by the graphs presented

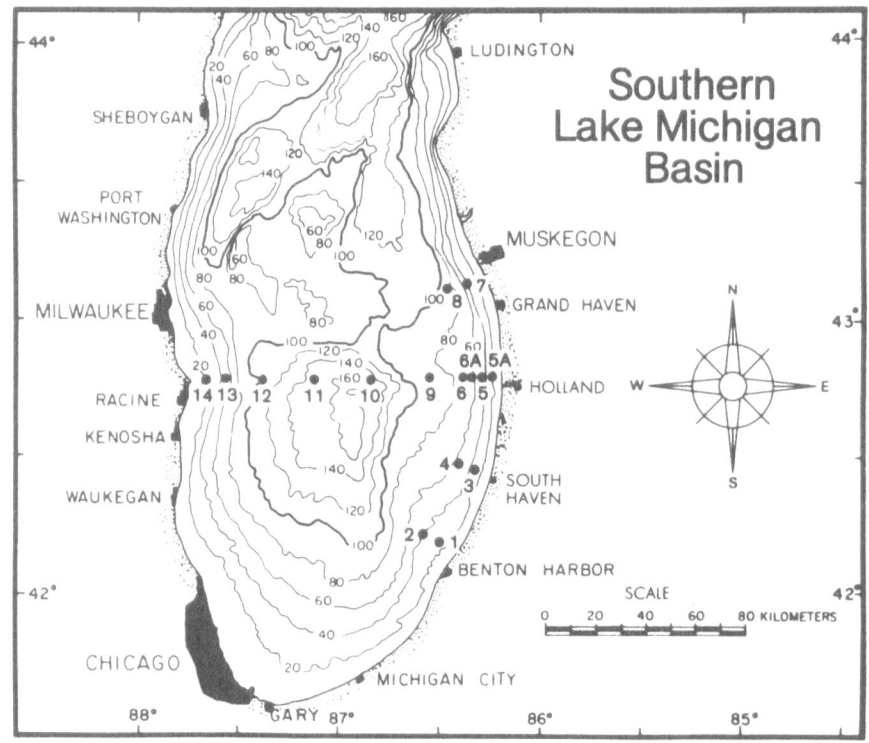

Figure 1.2a

Location map showing the bathymetry of Southern Lake Michigan
and the positions of the 16 current meter moorings.

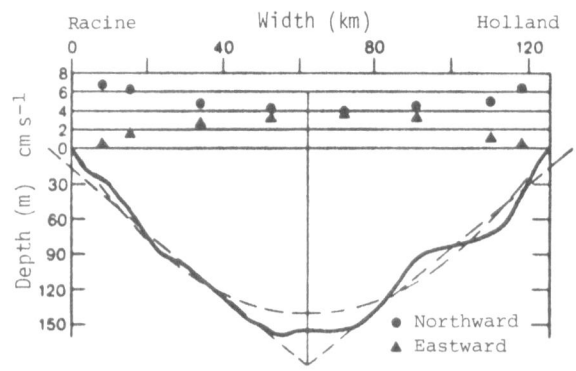

Figure 1.2b

Average amplitudes of the north and east
velocity components for rotational oscillations on the Southern Lake Michigan cross section during an episode of
wave excitation, 1-15 July 1976. Basin
approximations as a paraboloid or as an
inverted cone that are referred to later are also shown.

in *Figure 1.4*. They show a pro-
gressive vector diagram of the
hourly averaged low-pass filtered current velocities from
the 32 m depth current meter
at station 11 (*Figure 1.4a*) and
coherence and phase difference between the east and north
velocities from the same current meter (*Figure 1.4b*).

The hodograph clearly reveals
the oscillatory wave motion
with a period of about 4 days,
the rotation of the current
vector being in a cyclonic
(counterclockwise) direction

Spectral energy density $(cm\,s^{-1})^2\,cph^{-1}$ Spectral energy density $(cm\,s^{-1})^2\,cph^{-1}$

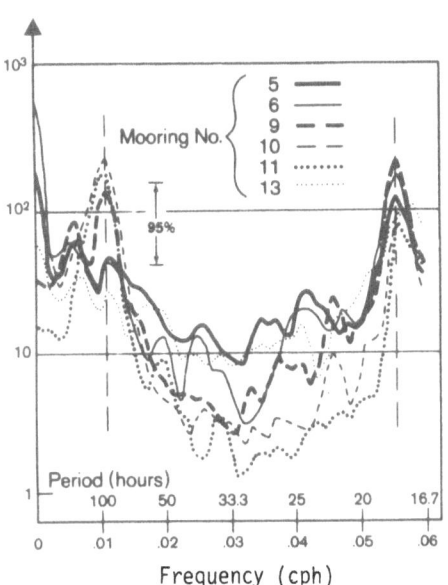 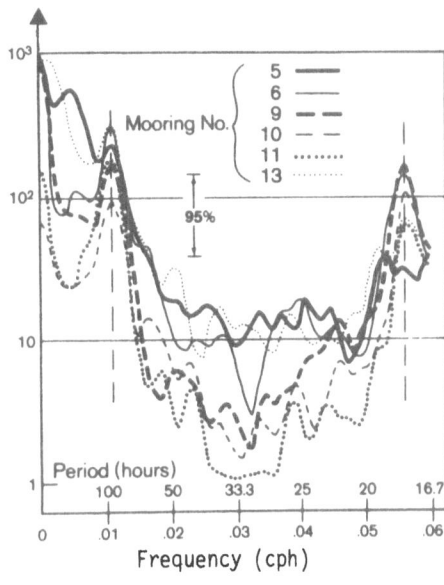

Figure 1.3a

Kinetic energy spectra of the eastward velocity component (u) recorded
at the 25 m level at six stations

Figure 1.3b

As in Figure 1.3a, except for the
northward velocity component (v).

in southern Lake Michigan. Vertical lines are drawn to accentuate the
two conspicuous energy peaks at near inertial and near 4-day periods.

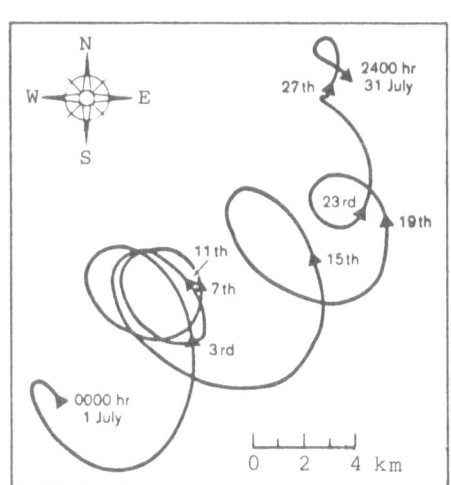 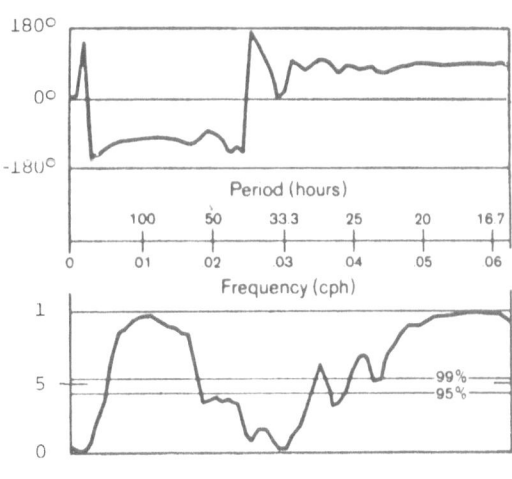

Figure 1.4a

Hodograph of hourly averaged, low-
pass filtered current velocities
from the 32 m depth current meter
at station 11 for the month of July
1976.

Figure 1.4b

Coherence (lower) and phase between
the east and north velocities at
the 32 m level from station 11. Two
broad areas of high coherence high-
light the cyclonic rotating 4-day
oscillation and the anti-cyclonic
near inertial oscillation.

[Figures 1.2 to 1.4 from Saylor, Huang & Reid (1980)]

for this station, which is in the center of the lake basin. Coherence and phase plots in *Figure 1.4b* for the East (u)- and North (v)-components for one current meter at position 11 accentuate this, as they show two highly correlated signals centered at the two indicated frequencies. At the 4-day period the u-component leads the v-component by a phase angle of 90° which corresponds to a cyclonic rotation; at the near-inertial period the phase angle of the u-component lags that of the v-component by approximately 90°, indicating an anti-cyclonic oscillation.

Further scrutiny of the data shows that (i) the 4-day period motions are clockwise rotations at all stations except 10 and 11 in the center of the lake, where they are counterclockwise and (ii) currents at on-shore positions are primarily along-shore while those at off-shore stations may have appreciable amplitudes for both, the East- and North-components.

These observations can be interpreted in terms of free TW's in a circular basin with nearly conical profiles. For a depth profile of the form

$$H = H_0\left(1-\left(\frac{r}{a}\right)^q\right), \quad q > 0 \tag{1.1}$$

(H_0 is the maximum depth at the center, a the radius of the circular basin and q a parameter characterizing the profile), Saylor, Huang & Reid (1980) deduce the frequency relation

$$\frac{f}{\omega} = \frac{3m + 2q}{m}. \tag{1.2}$$

Here ω is the frequency, f the Coriolis parameter and m = 1,2,3 ... the radial mode number. Values for the frequencies or periods are summarized in *Table 1.1* and the streamline pattern of the vertically integrated transport for the lowest two modes is sketched in *Figure 1.5*. The fundamental mode (m = 1) enjoys all properties of the observations mentioned above. In particular for the conical profile, its period is close to the observed 90 hour period. *Table 1.1* and (1.2) however, also show that for each mode the periods depend

		m = 1		m = 2		m = 3	
Profile	q	$\frac{f}{\omega}$	T [hours]	$\frac{f}{\omega}$	T [hours]	$\frac{f}{\omega}$	T [hours]
	1	5	84.5	4	67.6	3	50.7
	2	7	118	5	84.5	4	67.6
	3	9	152	6	101	5	84.5
	∞	0	∞	0	∞	0	∞

Table 1.1
Eigenfrequencies (-periods) of the first three TW-modes computed according to (1.2) using $f = 2\pi/16.9\,h$.

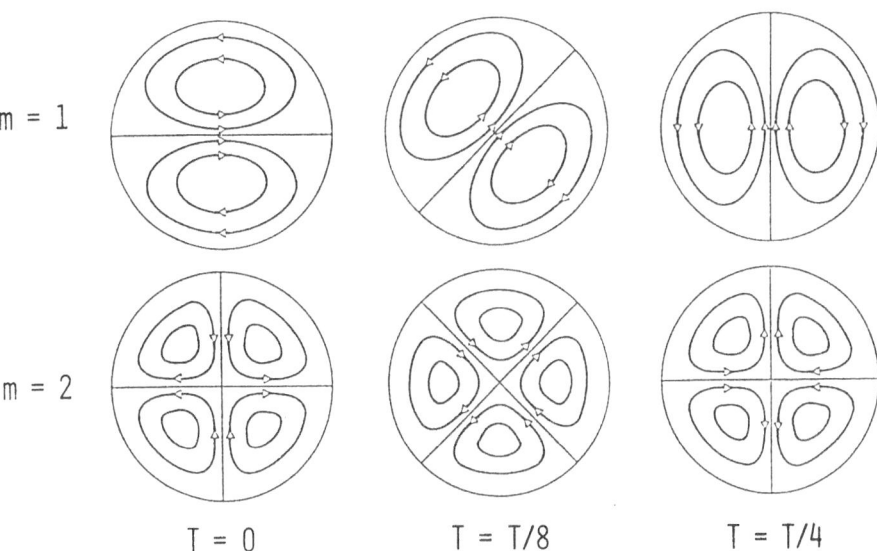

$$m = 1$$

$$m = 2$$

$$T = 0 \qquad T = T/8 \qquad T = T/4$$

Figure 1.5
Schematic sketch of the mass transport streamline pattern of the
fundamental and first "higher" mode plotted for three instances
during a quarter period. During one cycle, the system of gyres
rotates counterclockwise (on the Northern hemisphere) around the
basin.

strongly on the topography, but not on the size of the basin (H_0 and a do
not enter into the frequency relation). Moreover, the same period arises
for different modes and different bathymetries, indicating that the mode
structure is very important in interpreting observations. For details of
this model the reader is referred to section 3.2.

b) *Lake of Lugano (North basin)*

Southern Lake Michigan is a large and relatively shallow lake with a
width of approximately 100 km, whereas Lake of Lugano is small (17km long,
approximately 2 km wide and 300 m deep). According to current thinking,
existence of topographic waves should come as a surprise.

In summer 1979, thermistor chains and current meters were positioned at
various locations in the stratified Northern basin of Lake of Lugano.
Only temperature-time series could be analysed, [for a detailed description see Hutter, Salvadè & Schwab (1983); Mysak, Salvadè, Hutter & Scheiwiller (1983), (1985)]. They disclose a long-periodic signal which is
weaker than other oscillation signals, but nevertheless conspicuous enough
to be distinctly recognizable by eye in isotherm depth time series.

Figure 1.6 shows the map of Lake of Lugano (North basin) with indicated

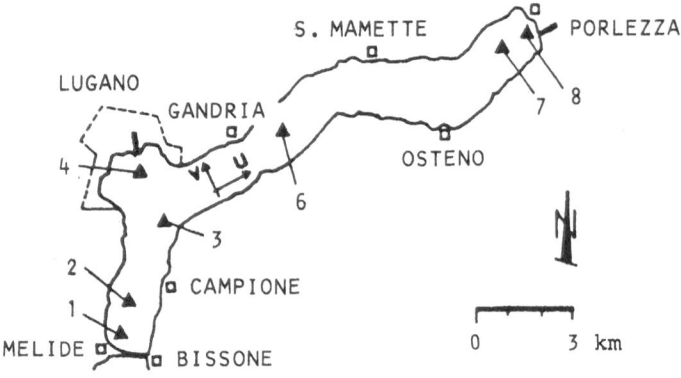

Figure 1.6   Northern Lake of Lugano with indicated
              positions of moorings.

stations where the wind and temperature (generally at the lower portion
of the metalimnion) were measured in July/August 1979. *Figure 1.7* summarizes wind data at stations 4 and 7 (top) and isotherm-depth-time series
at the stations 1 and 8, 6 and 4 (lower part). These time series "demonstrate a strong component of the motion with a period of perhaps 74 hours
(marked by circles ◐). To emphasize this wave the troughs of the isotherm
depths have been brought into prominence by thick solid and dotted lines.
The front arises first at station 4, propagates southwards, reaches station 1 approximately 7 hours later (indicated by the solid line connecting the troughs at Cassarate and Melide) and is reflected at the southern end of the lake. The reflected wave passes through station 4 again
(though split up into two smaller minima with an intermediate maximum,
indicated by an arrow marked with an encircled 1), and after 37 hours,
can be recognized at station 8 with a conspicuous minimum (heavy dotted
lines). It appears later at Cassarate, Melide and as a reflected wave at
Cassarate (arrow marked by an encircled 2), Porlezza, etc.", (Hutter, Salvadè & Schwab, 1983). The corresponding wave speed of approximately 12
cm s^{-1} is substantially lower than the wave speed of the internal gravity wave of the two-layer model; neither can it be explained as a higher order baroclinic Kelvin wave of, say, a three-layer model. Furthermore, *direct* wind forcing can be excluded as a likely cause of excitation (Mysak et al., 1985).

A careful statistical analysis of the temperature-time series shows that
the phase of the oscillation with a suspected 74-hour period propagates
counterclockwise around the basin.   *Figure 1.8* shows variance spectra,
coherence and phase difference of time-series pairs of the mean temperature displacement function for an event of 43 days duration. The energy

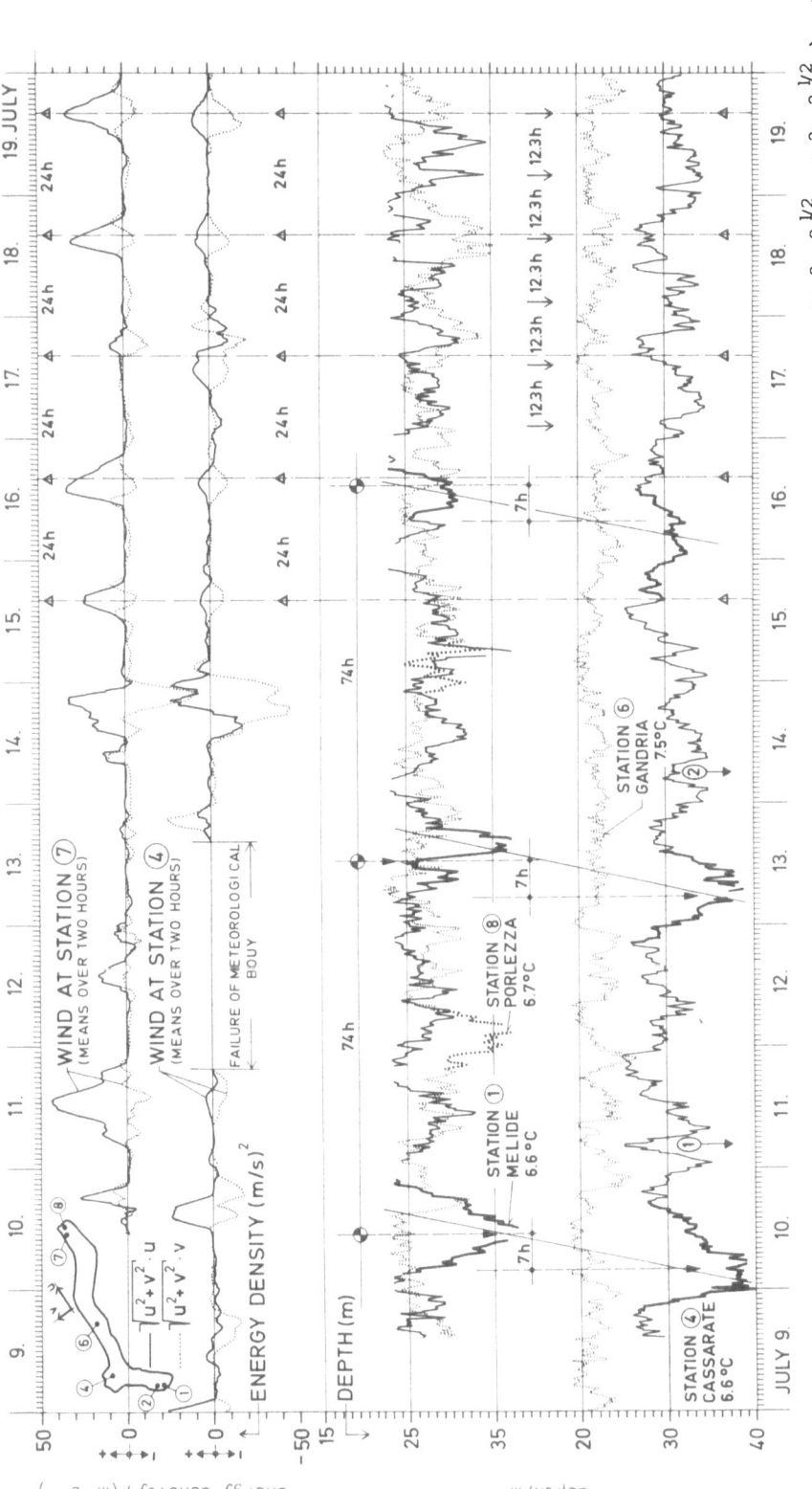

Figure 1.7 Time series of 2h means of the longitudinal and transverse components of the wind energy ($(u^2+v^2)^{1/2}u$, $(u^2+v^2)^{1/2}v$) with directions of u and v components as indicated in the insert in the upper left corner) at the stations 7 (Porlezza) and 4 (Cassarate) and unfiltered time series of selected isotherm depths at the stations 1 (Melide, top solid curve), 8 (Porlezza, dotted, superposed on the Melide curve) and 4 (Cassarate, solid curve). Components of the motion with conspicuous periods are marked by special symbols: troughs in heavy solid and dotted lines and marked by circles ● are for 74 h; triangles for 24 h; arrows for 12 h or 8 h. The time shown is from 9 to 19 July 1979. [From Hutter et al., 1983].

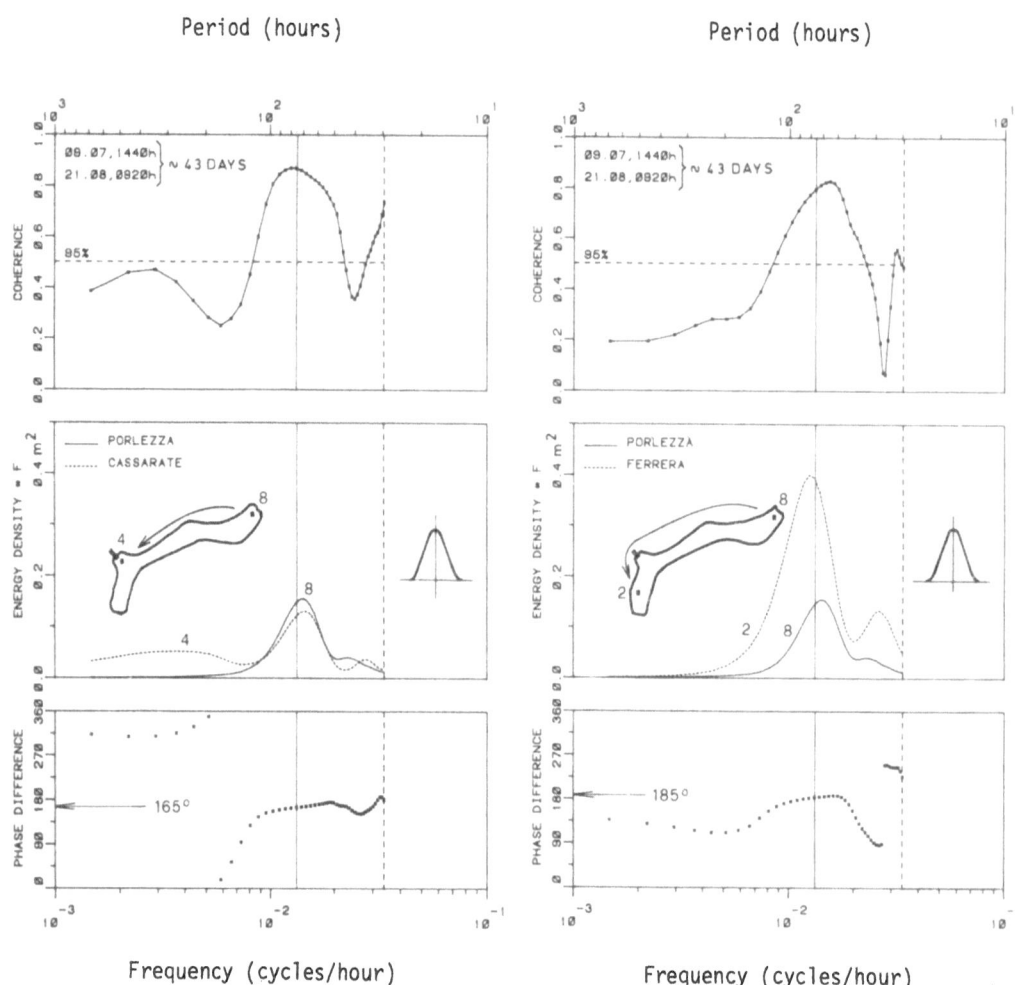

Figure 1.8 Variance spectra, coherence and phase difference of time-series pairs of the mean temperature-displacement function for an event of 43 days duration. Each column is subdivided into three subfigures where coherence c, variance and phase difference are plotted. The vertical broken line marks the cut-off period of 30h and the horizontal broken line in the coherence diagrams mark the 95 % confidence limit of arctan c. The insert maps show the station pairs, the arrow indicating the direction of phase progression. The inserted window indicates the tapering of the spectra. The vertical solid line marks the period of 74 h, and the arrow indicates the phase difference between the time-series pair at that period.

spectra of all mean temperature-displacement functions*) have a maximum between approximately 69 and 80 hours (with a mean of 74.5 hours), where coherence between data pairs is generally high.

The maps inserted in the graphs of the energy spectra show the locations of the stations where the temperature measurements from the thermistor

---

*) The mean temperature-displacement function is representative of a mean isotherm depth-time series. Its exact definition is not relevant here and can be looked up in Hutter et al. (1983).

cont. Figure 1.8

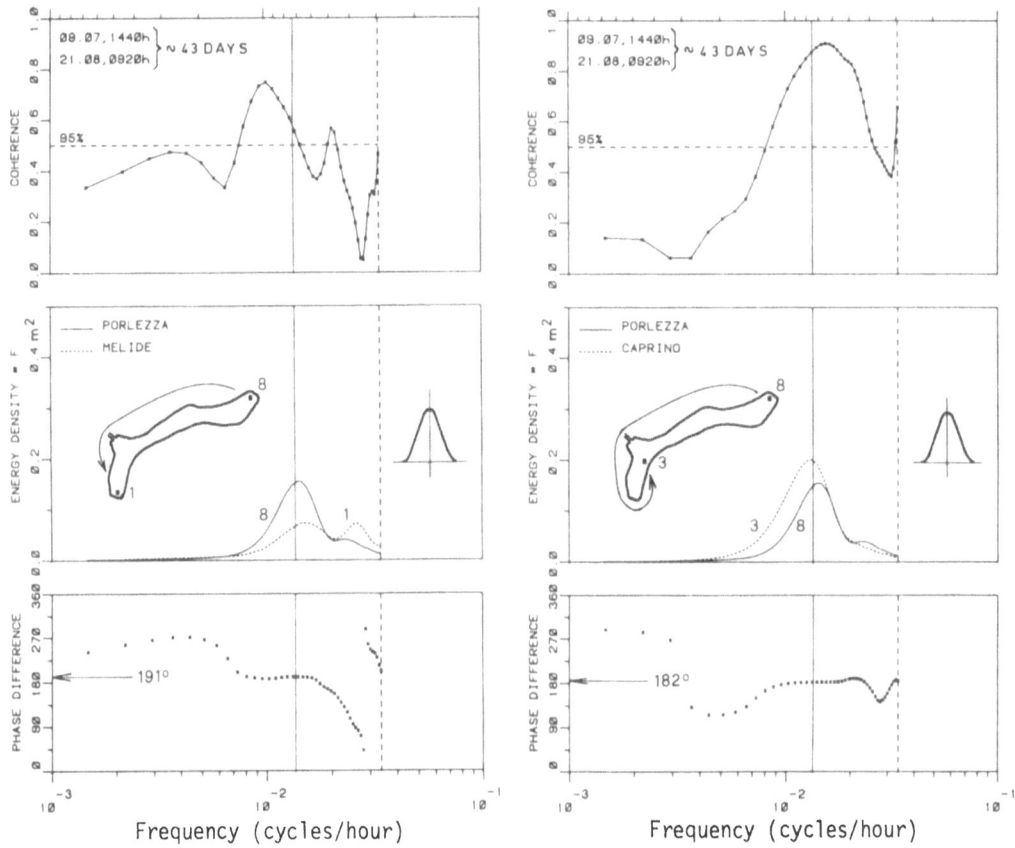

Frequency (cycles/hour)   Frequency (cycles/hour)

chains were taken, the arrow indicating the direction in which the phase propagates. Phase differences between station pairs at the 74 hour period are indicated in *Figure 1.8* by an arrow at the left margin of the corresponding phase difference.

Mysak et al. (1985) have explained the 74 h oscillation and the anticlockwise propagation of the phase as the baroclinic trace of a barotropically driven TW. They explain all their observations (except the "discrepancy" in the phase of the pair 3/8 of the mean temperature displacement function, which interrupts the anticlockwise increase of the phase difference) in terms of the fundamental mode of TW's in an elliptical two-layer basin. The streamlines as constructed by Johnson (1986) are sketched in *Figure 1.9*[*].

---

[*] *Mysak et al. (1985, p. 52) list six arguments in support of the TW-model and only the above mentioned discrepancy in the phase relation against it.*

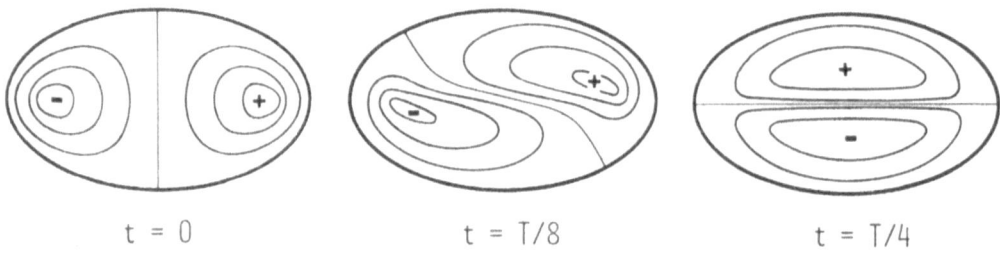

Figure 1.9   Contours of the mass transport stream function of the fundamental mode for the elliptical topographic waves [From Johnson (1986)].

To round out this picture it must be mentioned that the TW-equation has
been approximately and numerically solved by the finite element technique (Trösch, 1984) with results which do not at all support the interpretation using elliptical TW's.
Trösch finds that solutions in
the 65- to 95-hour-period range
are generally localized to the
two narrow ends and the bay of
Lugano. *Figure 1.10* shows the
streamline pattern of three such
modes, having periods of 68.5,
80.5 and 91 hours, respectively.

At the present stage we are not
able to favour one interpretation over the other. We can simply ask the questions:

(i) Is it an oversimplification
to approximate a complicated
real geometry (the Lake of
Lugano) by a mathematically
simpler (elliptical) basin
to which analytical TW-solutions can be constructed?

(ii) Can the finite element solution for Lake of Lugano be
trusted; i.e. are the discretized solutions for the Lake

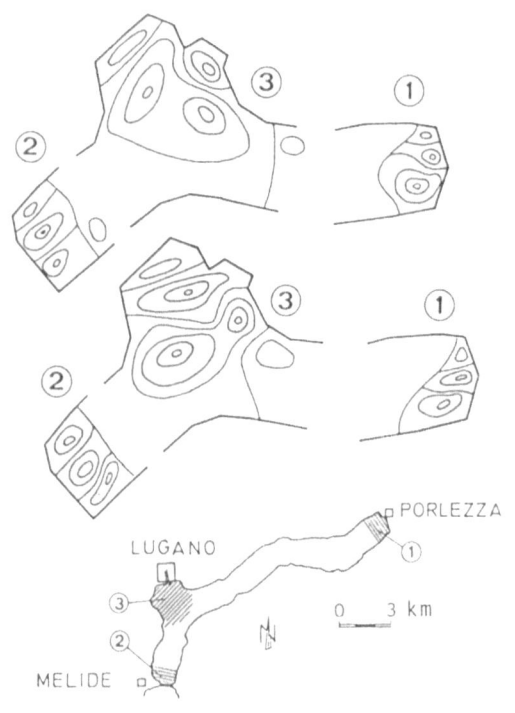

Figure 1.10

Three modes of longperiodic waves
in Lake of Lugano obtained by finite element technique.

$T_1 = 68.5$ h ⎫
$T_2 = 80.5$ h ⎬ for t = 0 (above) and
$T_3 = 91$ h ⎭

t = T/4   (below).

[From Trösch (1984) with alterations]

of Lugano adequate representations of the continuous TW-equation
for that lake?

This book will give some answers to these questions.

c) Lake of Zurich

Lake of Zurich (main basin) is 30 km long, 136 m deep and 2 to 3 km wide.
Its bathymetry is more complex than that of Lake of Lugano. Lake of Zurich consists of a relatively wide and shallow Eastern part between Rap-
perswil and Stäfa and a more narrow and deeper portion between Stäfa and
Zurich, Figure 1.11. During August/September 1978 current meters and
thermistor chains were deployed at the positions indicated in Figure 1.11.
The internal gravitational seiche response of this lake was analysed by
Mortimer & Horn (1982) and Horn, Mortimer & Schwab (1986). It revealed
oscillations with periods of 44 hours and smaller. The filtered long-periodic oscillations have not yet been studied in detail, but Figure 1.12

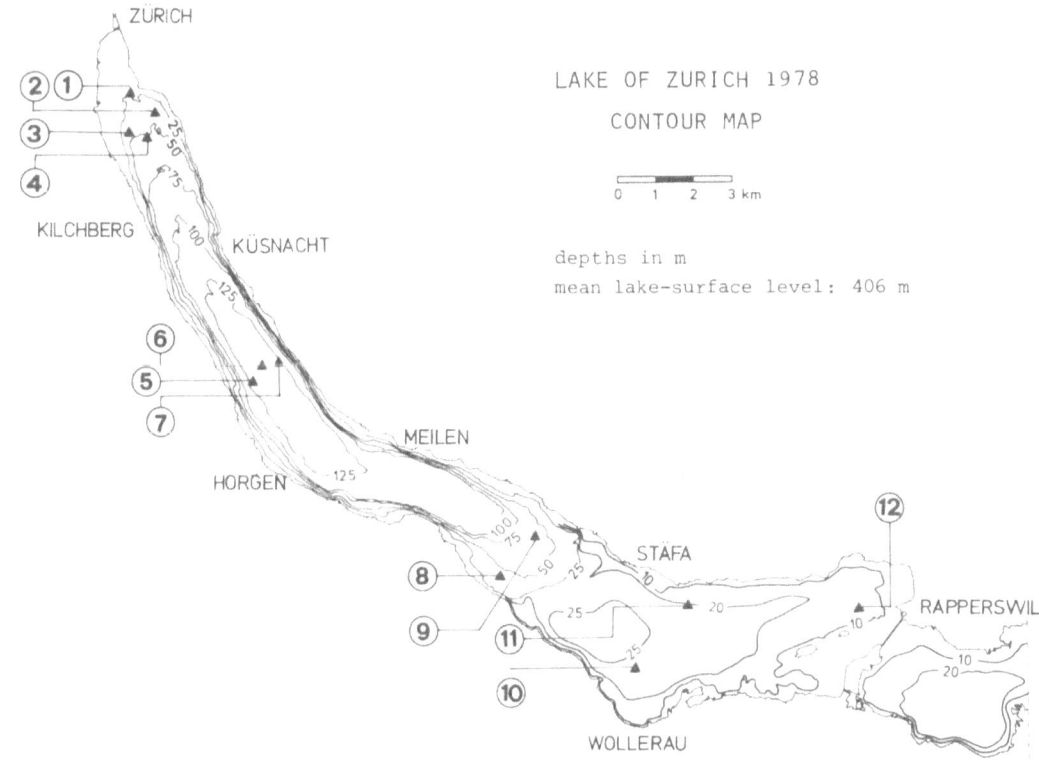

Figure 1.11

Bathymetric chart of the Lake of Zurich and positions of the 1978 mooring positions. [From Mortimer & Horn, 1982].

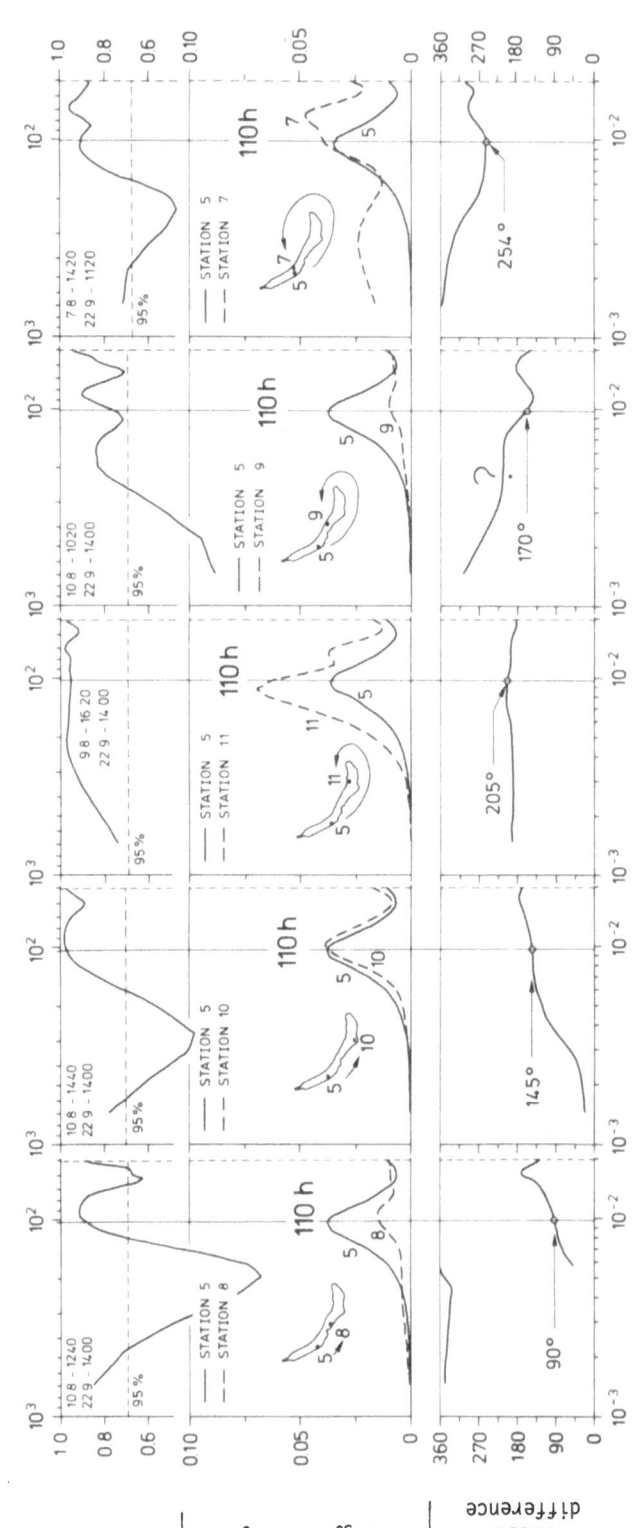

Figure 1.12

Coherence, variance and phase difference for the station pairs (5, 8),
(5,10), (5, 11), (5, 9), (5, 7), (5, 2), (5, 3), (5, 4), and (5, 6) in the low
frequency range for the 10°C-isotherm-depth-time series in the Lake of
Zurich experiment (August–September 1978). Energy spectra peak systematically at 100–110 hours and, with one exception phase propagation is
counterclockwise. [From Hutter & Vischer (1986)].

cont. ⟶

cont.
Figure 1.12

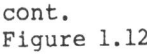

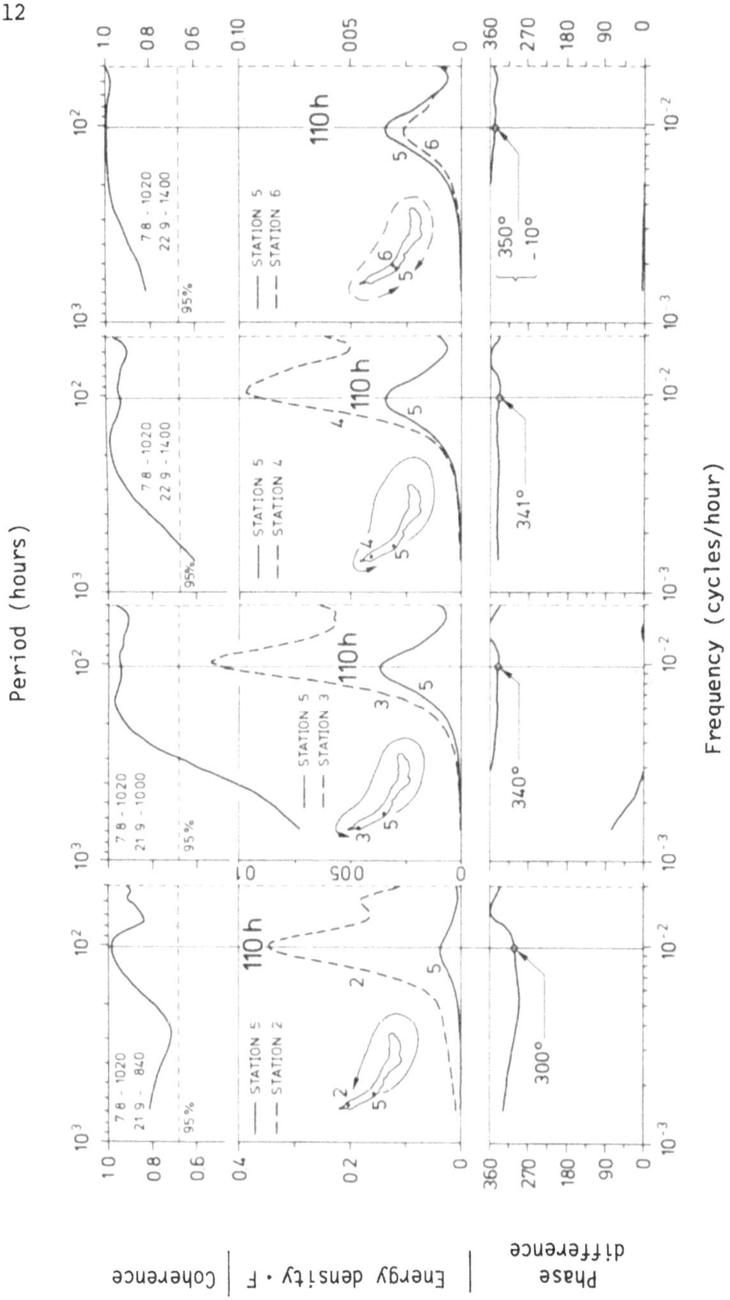

shows coherence, variance and phase difference for the 10°C isotherm-
depth-time series for the station pairs indicated in the insert maps.
Energy spectra peak systematically at 100-110 hours with uniformly high
coherence, and the phase at the 110-hour period propagates (with one exception, station pair 5 and 9) in the counterclockwise direction. Furthermore, rotary spectra of the water current at the positions and depths

as indicated in *Figure 1.13* show that near the 100 h period

(i) the rotation of the current vector at the mid-lake position 6 (approximately deepest point) is predominantly counterclockwise,

(ii) it is predominantly clockwise at the near shore position 7 but shows no clear distinction in direction at positions 4 and 12.

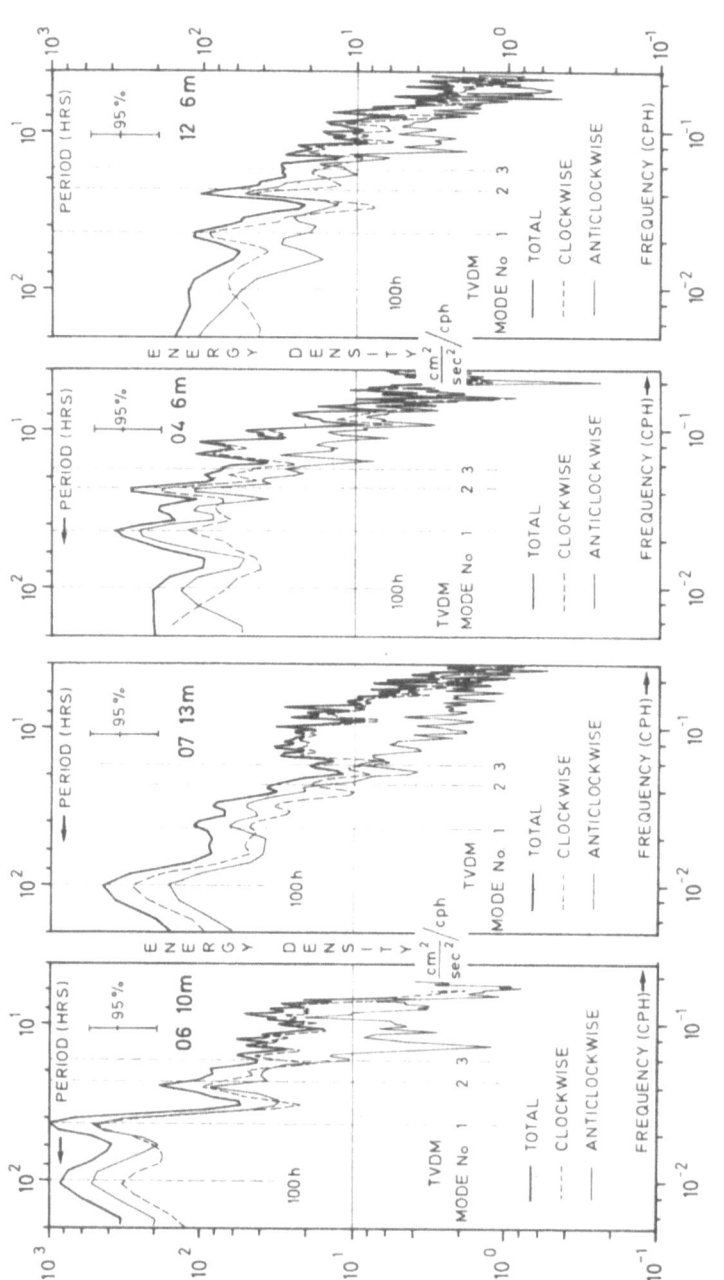

Figure 1.13

Rotary spectra of current at the following moorings and depths shown, from left to right: 6(10m), 7(13m), 4(6m) and 12(6m). Total energy is indicated by the upper thick continuous line, while the clockwise-turning and anticlockwise-turning contributions are indicated by the dashed and continuous thin lines, respectively.   [From Horn, Mortimer & Schwab (1986)].

The statistical confidence of these results is, however, not very strong (compare confidence intervals in *Figure 1.13*).

These results at least suggest an interpretation in terms of a basin wide TW-mode, and indeed all observations at the 100-110 hour period except the phase relations of the 10°C-isotherm-depth-time series at station 9 are in conformity with the structure of the fundamental mode of an elliptical TW. Mysak (1985) takes this view. On the other hand, numerical finite difference or finite element solutions have so far not been constructed for TW's in the Lake of Zurich, but given the conflicting results of Lake of Lugano a cautious attitude is advisable.

d) *Lake Ontario*

This lake has the longer history of TW-modelling than the Swiss lakes, but available data from current meter and/or temperature recordings do not seem to be detailed enough to permit clear inferences regarding basin-wide TW-response. Episodic informations are available, and the behavior near the shores is better documented.

The coastal strip model of Csanady (1976) is based on Gill & Schumann's (1974) shelf wave analysis. For the idealized inclined plane beach-coastal strip model (*Figure 1.14a*), using a shore zone of 20 km width, this model produces a phase propagation of the TW of 0.5 m s^{-1} and a distribution of the amplitude of the long-shore velocity as indicated in *Figure 1.14b*.

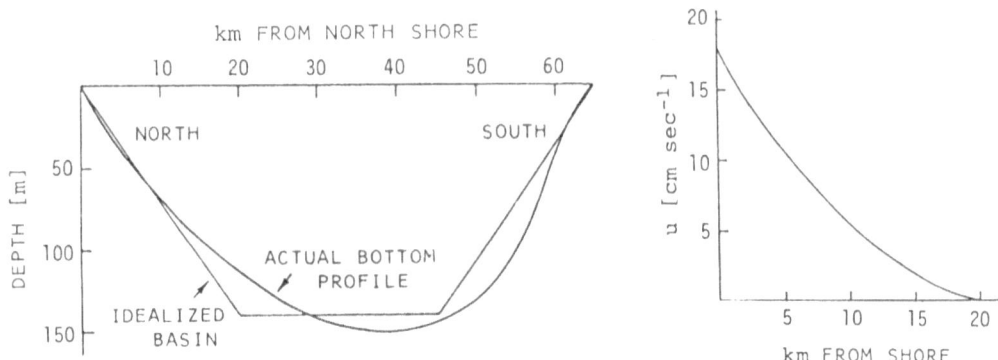

Figure 1.14a

Actual and idealized depth distribution in Lake Ontario along Oshawa-Olcott cross section.

Figure 1.14b

Properties of topographic wave along north shore of idealized basin: long-shore velocity, for an arbitrary shore elevation amplitude of 1 cm.

Detailed coastal-zone observations were carried out along the North shore of Lake Ontario during IFYGL (International Field Year for the Great Lakes), at Oshawa and Presqu'ile. Following Eastward wind impulses, a TW may be expected to propagate Eastward along the Northern shore from Presqu'ile to Oshawa in 2 or 3 days. Such an event (5 days after a massive Eastward wind impuls) is shown in *Figure 1.15*. The velocities are all Westward (leaving the shore to the right) and peak at 0.3 ms^{-1}. The slower flow along the shore is due to friction and that at the top layer is of baroclinic origin. The current amplitude decayed appreciably at 12-14 km from shore (not shown in Figure 1.15).

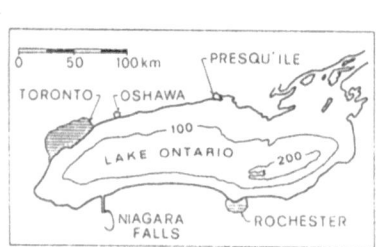

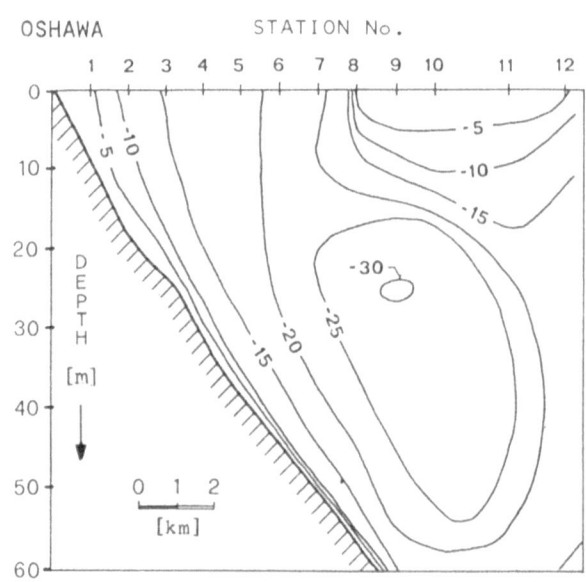

OSHAWA   STATION No.

Figure 1.15

Longshore velocity (cm s^{-1}) in a cross-section off Oshawa, Ontario, October 13, 1972. Negative values designate Westward velocity, while previous wind impulse was Eastward. [From Csanady (1976)].

This coastal strip model cannot explain the typical topographic gyre pattern, see *Figures 1.5* and *1.9* as no return flow in the offshore regions is modelled by it. A channel model, or closed basin model must account for this. Hamblin (1972) presents finite element calculations of the third azimuthal mode of an elliptic paraboloid model of Lake Ontario and produces a period of 10 days and Simons (1983) analyses the resonant topographic response of nearshore currents to longshore winds and finds that Hamblin's mode is likely to be excited by the winds. Saylor and Miller (1983) report that rotary spectra computed for current moorings near the center of Lake Ontario disclose no unequal partitioning of energy in clockwise and anticlockwise components.

On the other hand, waves with periods of 4 to 5 and 10 to 25 days and with clockwise rotating current vectors at near-shore moorings were found to propagate cyclonically about the coast by Marmorino (1979). *Figure 1.16b* shows the squared coherence between hourly alongshore currents at the stations shown in *Figure 1.16a*. Marmorino concludes that the 10 to 25 day signal may represent free TW's while the shorter signal, having a period near the peak in the wind spectra may be the result of forced motion. However, near resonance behavior of another TW is equally plausible. We conclude that the long periodic response of Lake Ontario may be a mixture of several free TW's that are excited by the wind.

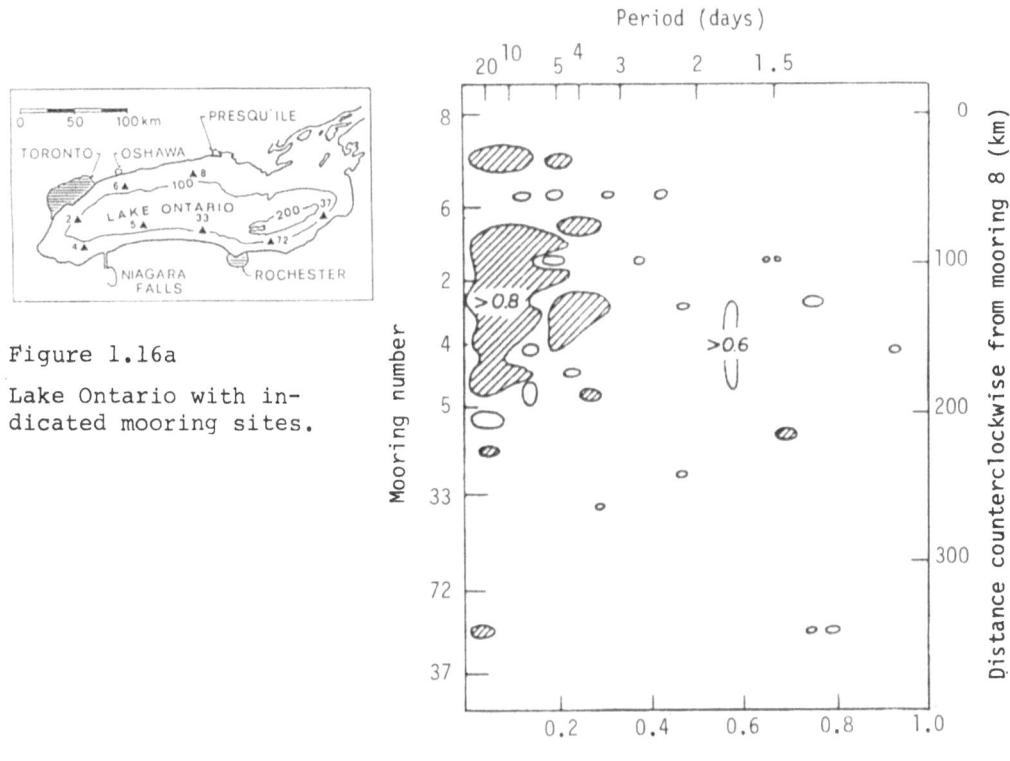

Figure 1.16a

Lake Ontario with in-
dicated mooring sites.

Figure 1.16b

Frequency (cpd)

The squared coherence between hourly alongshore currents at different stations plotted against frequency and average location. Five raw spectral estimates were averaged, yielding about 10 degrees of freedom. Only values greater than the 95 % significance level (0.6) are shown. The hatched areas represent values greater than 0.8: the other contour lines enclose values of the squared coherence between 0.6 and 0.8. [From Marmorino (1979)].

e) *Other lakes and ocean basins*

There is ample further evidence of vorticity generated flow features reminiscent of TW-behavior. Saylor & Miller (1983) also analyse time series of water currents from instruments moored in Lakes *Erie* and *Huron* at several offshore positions. They find for these lakes that in the period band of 85 to 125 hours kinetic energy associated with anticlockwise (clockwise) rotation of the current vector is accumulated for "mid-lake" (near-shore) mooring stations, suggesting a fundamental TW-mode as indicated in *Figure 1.5*. Given the limited observational evidence such a conclusion seems to be premature, however.

Numerical studies of the wind-induced circulation in the *Baltic Sea* (Simons, 1978, 1980; Kielmann & Simons, 1984) and interpretation of infrared satellite imagery (Kielmann, 1983), indicate local current patterns with cyclonically rotating gyres that are reminiscent of TW's induced by winds. *Figure 1.17* shows the observed and computed vertically integrated

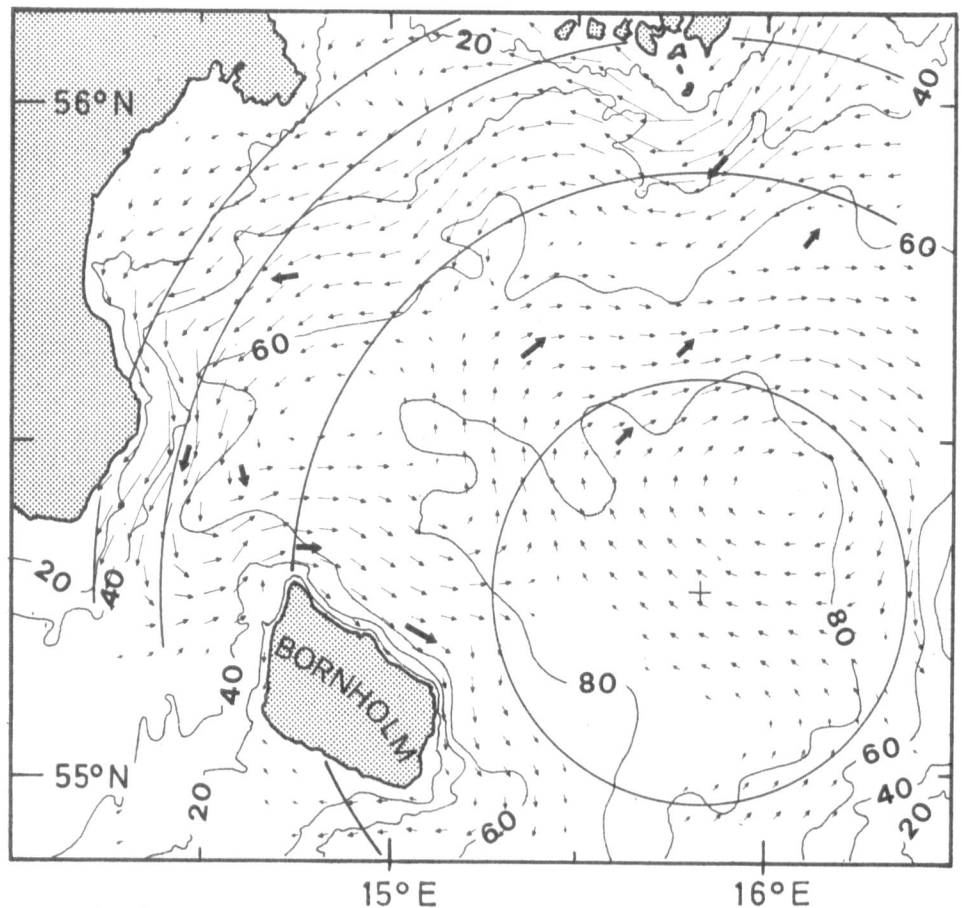

Figure 1.17

Observed (dark arrows) and computed (light arrows) vertically integrated transports in the South-West Baltic Sea after a strong wind from the Northeast, i.e. from the upper right-hand corner.
[From Simons (1978a), with additions].

transport in the *Bornholm basin* located in the South-West Baltic Sea after
a strong wind from the Northwest. This current pattern was established
from a configuration of completely reversed flow before the onset of the
wind, thus confirming the basic concept of large-scale vorticity generation by the interaction of wind stress and bottom slope. The figure
also shows the topography and its approximation by a circular basin.
Adopting this circular basin, Wenzel (1978) demonstrates that there are
good reasons to assume that the system of gyres is an excited higher
mode of TW's with a period of approximately 5 days. Kielmann (1983) reaches similar conclusions.

Similar vortex systems for the vertically integrated transport of the
Gulf of Bothnia are also attributed to wind generated topographic response
(Kielmann & Simons, 1984). Bäuerle (1984) models this gulf as a channel
having a trench profile; he constructs numerical TW solutions having
periods of 70 to 75 hours, but abstains from a comparison of his results
with observations.

We may conclude this brief overview with the following slightly simplified statements:

(1) There exists a large amount of episodic and isolated observational
    evidence which suggests that long periodic oscillating responses in
    lakes and local areas of such basins *may* be explained in terms of
    TW-models.

(2) Coherent temperature and current data covering an entire basin for
    a period of longer duration do at present not seem to be available
    in order to clearly identify the primary cause of the motion and to
    interpret the observations uniquely in terms of a model.

(3) It appears that long periodic circulation features which are the im-
    mediate result of strong winds, can be explained by simple idealized
    or more realistic and complicated models and both yield very similar
    if not identical results. Long periodic features which are the direct
    result of a strong wind gust permit interpretation in terms of TW's.

(4) On the other hand, inferences from simple (analytically accessible)
    models and more realistic (only numerically exploitable) models which
    attempt to explain basin-wide TW-behavior are so far conflicting.
    Hence, interpretation of basin-wide long periodic oscillations re-
    mains an open problem, at least as long as one cannot assume with cer-
    tainty that the numerically discretized models generate flows which

are the approximation (in a well defined sense) of the original non-discretized TW-equation to the real topography.

1.4 Aim of this work

The above summary makes it clear that our knowledge of TW-behavior is not sufficiently precise that water oscillations in lakes with periods of several days could be attributed with certainty to this type of wave motion. Our goal here is:

(i)   to review the existing literature on TW's,

(ii) to collect some analytical solutions of the TW-equation and to describe their properties,

(iii) to present a selection of numerical results of the TW-equation and to describe the difficulties in the construction of these, and

(iv) to present aporoximate methods by which a reliable description of the physics is obtained.

Emphasis is on the mathematical and computational side but always having the interpretation of observations in mind.

To indicate the present difficulties we mention that analytical solutions for the TW-equation were found so far only for very special basin shapes and topographies, such as the parabolic circles or ellipses of Lamb (1932) and Ball (1965) or the elliptic basin with exponential depth profile of Mysak (1984, 1985) and Johnson (1986); all these models could describe the periods of observed wave motion provided that a set of parameters was well chosen. It turned out, however, that the configurations, determined by the fitted parameters, did not show much similarity with the natural basins or profiles. For instance, in order to explain a topographic mode in Lake of Lugano Mysak et al. (1985) were forced to choose a basin which was much fatter than that of the Lake of Lugano and did not resemble its shape. They were not able to obtain the required value of the period when selecting realistic length to width ratios. It appeared that the topography at the long ends had to be reproduced in the model quite accurately, while the shoulders along the long steep shores were relatively unimportant; at least to some TW-modellers this is counterintuitive. Therefore, at the moment there is certainly a lack of adequate theoretical models which could satisfactorily explain topographic waves in realistic basins.

It is only natural to suppose that construction of numerical solutions to realistic basins will lead us out of the difficulty. However, as Bennett & Schwab (1981) have shown us for an elliptical basin with parabolic profile, finite difference solutions of the TW-equation may (for the lowest mode) or may not (for higher order modes) adequately approximate known analytical solutions. So, one ought to be very cautious when constructing numerical solutions to the TW-equation in enclosed basins[*].

Numerical solution techniques, on the other hand are effective tools in explaining wave motions under very specific aspects. However, they are not likely to enlarge our knowledge very much from a *physical* point of view. Furthermore, they require immense computational effort for a problem which, in its basics, still lacks a thorough understanding. Questions concerning the spectrum and the modal structure or behavior in curved and more complicated basins are still unanswered.

This is, why this monograph also proposes a method that treats the problem from another point of view. The problem is simplified by making assumptions which are apt to the specific situation of second class waves. We attempt to show a way which can combine the preciseness of an exact model with the facility of the basin modelling supplied by FD- or FE-methods. A test of such a model is likely to bring about also the important properties of a lake basin under the aspect of long-periodic wave motion. Furthermore, we obtain an answer about the quality and strength of such a method and a hint towards specific improvements of assumptions, which are invoked in the course of developments.

Many results will be constructed for infinite channels, but we have obtained results also for the reflection of TW's at a wall of a semi-infinite channel. The study of this reflection for varying topography close to the wall will shed light on the return-flow mechanisms of coastally trapped vorticity dominated currents and thus will path the routes to a more complete understanding of basin-wide TW-structures. The study also includes the presentation of TW's in enclosed basins which are obtained by adopting our channel method. These solutions will provide detailed insight into the physics of localized and basin-wide mode structures of TW's.

---

[*] *This must apply to all previously constructed finite difference and finite element solutions of the eigenvalue problem: eg. Hamblin (1972), Rao & Schwab (1976), Trösch (1984).*

2. Governing equations

In this chapter we introduce the equations and approximations from which the TW-equation and associated boundary conditions are derived. No completeness is intended.

2.1 Equations of adiabatic fluid flow

Fundamental to the description of the motion of water are the balance laws of mass, momentum and energy that can be applied to a fluid body. These three fundamental laws and the equation of state lead to the equations, quoting Pedlosky (1982),

$$\frac{\partial \underset{\sim}{u}}{\partial t} + (\underset{\sim}{u} \, grad) \, \underset{\sim}{u} + 2\underset{\sim}{\Omega} \times \underset{\sim}{u} = \underset{\sim}{g} - \underset{\sim}{\Omega} \times (\underset{\sim}{\Omega} \times \underset{\sim}{r}) - \frac{1}{\rho} \, grad \, p,$$

$$\frac{\partial \rho}{\partial t} + div \, (\rho \, \underset{\sim}{u}) = 0,$$

$$\frac{\partial \rho}{\partial t} + (\underset{\sim}{u} \, grad) \, \rho = 0, \tag{2.1}$$

$$\rho = \rho \, (p, T),$$

in which the chemical and viscous aspects of the problem have been ignored. The system (2.1) describes adiabatic fluid[*) motion in a system subject to steady rotation; in other words, (2.1) contains all aspects of geophysical fluid dynamics. Mathematically, (2.1) constitute five nonlinear partial differential equations and one algebraic equation, the equation of state. Complemented by appropriate boundary conditions these six equations determine the six unknown fields

$$\begin{array}{ll} \underset{\sim}{u}(\underset{\sim}{x}, t) & \text{velocity field,} \\ \rho(\underset{\sim}{x}, t) & \text{density field,} \\ p(\underset{\sim}{x}, t) & \text{pressure field,} \\ T(\underset{\sim}{x}, t) & \text{temperature field,} \end{array}$$

which are all functions of space and time. The given fields are

$$\begin{array}{ll} \underset{\sim}{\Omega}(\underset{\sim}{x}) & \text{angular velocity[**)} \\ \underset{\sim}{g}(\underset{\sim}{x}) & \text{gravity field.} \end{array}$$

---

*) A process is called adiabatic, if mass and heat diffusion are ignored. Thus, if also radiative and shear heating are neglected, the balance of internal energy implies $d\varepsilon/dt = 0$, which can be expressed as $d\rho/dt = 0$.

**) On the f-, or β-plane (a plane tangential to the globe) with the x-axis pointing towards East, y-axis pointing towards North and the z-axis pointing in the radial direction $\underset{\sim}{\Omega}$ has the components $\underset{\sim}{\Omega} = (0, \tilde{f}/2, f/2)$ where

$$f = 2 \, |\underset{\sim}{\Omega}| \, \sin \phi, \qquad \tilde{f} = 2 \, |\underset{\sim}{\Omega}| \, \cos \phi.$$

ϕ is the latitude angle and f the Coriolis parameter. We assume here mid latitude positions on the Northern hemisphere for which case $\phi > 0$ and $f \approx 10^{-4} \, s^{-1}$. We shall also assume f to be constant ($\beta \equiv \partial f/\partial y = 0$, no β-effect).

Equations (2.1) are completed by boundary and initial conditions. The boundary conditions are:

(i) at the lake bottom $F_B \equiv H(\underset{\sim}{x}) + z = 0$, a tangency condition of the flow and the prescription of the heat flow through the bottom (usually assumed to be zero):

$$\left. \begin{aligned} \underset{\sim}{u} \cdot \text{grad } F_B &= 0, \\ \text{grad } T \cdot \text{grad } F_B &= 0, \end{aligned} \right\} \text{ at } F_B(\underset{\sim}{x}) = 0, \tag{2.2}$$

(ii) at the free surface $F_S \equiv \zeta(\underset{\sim}{x}, t) - z = 0$, a kinematic condition, the prescription of the atmospheric stresses $\underset{\sim}{t}_S^{atm}$ and the prescription of the surface temperature:

$$\left. \begin{aligned} \frac{dF_S}{dt} &= 0, \\ \underset{\sim}{\sigma} \cdot \underset{\sim}{n}(\underset{\sim}{x}, t) &= \underset{\sim}{t}_S^{atm}(\underset{\sim}{x}, t), \\ T(\underset{\sim}{x}, t) &= T_S(\underset{\sim}{x}, t), \end{aligned} \right\} \text{ at } F_S(\underset{\sim}{x}, t) = 0. \tag{2.3}$$

In (2.2) and (2.3) $\underset{\sim}{n}$ is the unit normal vector, $\underset{\sim}{\sigma}$ is the stress tensor in the fluid and $\underset{\sim}{t}_S^{atm}$, T_S are known functions of position. The boundary conditions represent constraints on the motion, in that they select for instance *eigenfrequencies* of seiches and "quantize" the otherwise free waves in closed basins. Moreover, equations (2.1)-(2.3) pertain to a broad spectrum of wave motion (external and internal waves): gravity waves or seiches, Kelvin waves, Poincaré waves, shelf waves, topographic waves, etc. Not only water motion on the Earth but equally atmospheric motion can be explained. Among these are buoyancy waves, Föhn waves, frontal motions, Rossby waves, etc.

Parallel with the generality of the above equations goes the difficulty to solve them. A general solution, which would embrace all aspects of fluid motion in a given configuration (e.g. channel, lake basin, atmosphere, etc.) is not yet found and is not worth searching for. The alternative is to introduce more or less reasonable neglections and approximations which (i) simplify the system (2.1)-(2.3) (ii) filter out all those effects which are not of concern but (iii) retain the characteristics of the motion of interest. This approximation procedure has cast light on various different domains of the spectrum. These often lie apart and form distinct regimes with distinct behaviors. Connections to overlapping mechanisms can sometimes be obtained by adopting perturbation analyses.

2.2 Vorticity, potential vorticity, topographic Rossby waves

In order to derive the governing equation of TW's consider the conservation of a quantity called *potential vorticity*. Let us demonstrate here by means of a simple model that the equation of topographic Rossby waves emerges essentially from the conservation of *angular momentum*. In a second step, by deriving the evolution equation of potential vorticity, this statement will be put into a more rigorous setting.

As a preparation to the model we need the definition of the (relative) *vorticity* $\omega \equiv \text{curl } \underset{\sim}{u}$ of the velocity field $\underset{\sim}{u}$. Recall that for a rigid rotation with angular velocity Ω^R this vorticity is $2\,\Omega^R$. Hence, a rigid rotation with angular velocity Ω^R relative to the f-plane rotating with Ω has an absolute angular velocity $\Omega^A = \Omega^R + \Omega$ and an absolute vorticity which is twice this value.

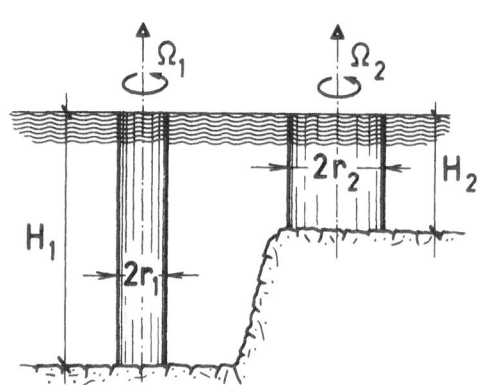

Consider the simple model sketched in *Figure 2.1*. Isolate a water column, which is assumed to be a rigid cylinder rotating about its vertical axis. The angular velocities in the two respective positions are Ω_1 and Ω_2. The angular velocity, which column 1 will take upon transportation to position 2 can be calculated when the conservation laws of mass and angular momentum are applied.

Figure 2.1

A mechanical analogy of the mechanism of the topographic wave motion.

Balance of mass in the columns requires

$$m/(\rho\pi) = r_1^2\, H_1 = r_2^2\, H_2 \tag{2.4}$$

and conservation of angular momentum yiels

$$\tfrac{1}{2}\, m\, r_1^2\, \Omega_1 = \tfrac{1}{2}\, m\, r_2^2\, \Omega_2. \tag{2.5}$$

Equations (2.4) and (2.5) are satisfied provided the quantity Ω/H following the fluid motion remains constant:

$$\Omega/H = \text{constant.} \tag{2.6}$$

Because the vertical component of the absolute vorticity of a rigid body motion is twice the *total* angular velocity, equation (2.6), on the rotating Earth, is tantamount to the statement

$$(\omega_z + f)/H = \text{constant,} \tag{2.7}$$

where ω_z is the vertical component of relative vorticity and f the Coriolis parameter. This quantity must therefore be conserved when one
follows the fluid motion, implying that (2.7) takes the form

$$\frac{d}{dt}\left(\frac{\omega_z+f}{H}\right) = 0,\qquad(2.8)$$

which is the conservation law of *barotropic potential vorticity*. The operator d/dt is the convective derivative operator

$$\frac{d}{dt} = \frac{\partial}{\partial t} + \underset{\sim}{u}\,\text{grad},$$

in which $\underset{\sim}{u}$ is the fluid velocity.

To derive (2.8) more rigorously we form the curl of the momentum equation $(2.1)_1$. After fairly routine transformations (see e.g. Hutter, 1984a,
p. 25) this yields

$$\frac{d\underset{\sim}{\omega}_a}{dt} = \text{grad}\,\underset{\sim}{u}\cdot\underset{\sim}{\omega}_a - \underset{\sim}{\omega}_a\,\text{div}\,\underset{\sim}{u} + \frac{\text{grad}\,\rho \times \text{grad}\,p}{\rho^2}\qquad(2.9)$$

where

$$\underset{\sim}{\omega}_a = \underset{\sim}{\omega} + 2\underset{\sim}{\Omega} = \text{curl}\,\underset{\sim}{u} + 2\underset{\sim}{\Omega}.\qquad(2.10)$$

Accordingly, the material rate of change of the *absolute vorticity* $\underset{\sim}{\omega}_a$ is
made up of the three terms on the right hand side of (2.9). The last
term is the production of vorticity due to the fact that density gradients and pressure gradients are not parallel; this is vorticity production by baroclinicity. The first and second term on the right describe
the production of vorticity due to vortex tilting and vortex stretching
(see Pedlosky, 1982).

More useful than the concept of vorticity is the concept of *potential vorticity*, which was introduced by Ertel (1942). In the presentation below,
we follow essentially Pedlosky (1982).

To introduce the potential vorticity, replace in (2.9) div $\underset{\sim}{u}$ by $-\dot{\rho}/\rho$ to
write it in the form

$$\frac{d}{dt}\left(\frac{\underset{\sim}{\omega}_a}{\rho}\right) = \frac{1}{\rho}\,\text{grad}\,\underset{\sim}{u}\cdot\underset{\sim}{\omega}_a + \frac{\text{grad}\,\rho \times \text{grad}\,p}{\rho^3}.\qquad(2.11)$$

Consider a scalar quantity λ which satisfies the balance statement

$$\frac{d\lambda}{dt} = \psi_\lambda,\qquad(2.12)$$

where ψ_λ may incorporate flux, supply and production terms. Let grad λ
be the gradient field of λ and form the inner product of (2.11) with
grad λ. This yields

$$\text{grad}\,\lambda \cdot \frac{d}{dt}\left(\frac{\underset{\sim}{\omega}_a}{\rho}\right) = \left(\frac{1}{\rho}\,\text{grad}\,\underset{\sim}{u}\cdot\underset{\sim}{\omega}_a\right)\cdot\text{grad}\,\lambda + \text{grad}\,\lambda\cdot\frac{\text{grad}\,\rho \times \text{grad}\,p}{\rho^3}.$$

If one adds the identity

$$\frac{\underset{\sim}{\omega}_a}{\rho} \cdot \frac{d}{dt}(\text{grad}\,\lambda) \equiv \frac{\underset{\sim}{\omega}_a}{\rho} \cdot \text{grad}\,(\frac{d\lambda}{dt}) - \text{grad}\,\lambda \cdot (\text{grad}\,\underset{\sim}{u} \cdot \frac{\underset{\sim}{\omega}_a}{\rho})$$

and uses (2.12) these two equations combine to yield

$$\frac{d}{dt}(\frac{\underset{\sim}{\omega}_a \cdot \text{grad}\,\lambda}{\rho}) = \frac{\underset{\sim}{\omega}_a \cdot \text{grad}\,\psi_\lambda}{\rho} + \text{grad}\,\lambda \cdot \frac{\text{grad}\,\rho \times \text{grad}\,p}{\rho^3}. \qquad (2.13)$$

The quantity

$$\boxed{\pi_\lambda = \frac{\underset{\sim}{\omega} + 2\underset{\sim}{\Omega}}{\rho} \cdot \text{grad}\,\lambda} \qquad (2.14)$$

is called the *potential vorticity*. It follows that if

(i) λ is conserved for each fluid particle i.e., $\psi_\lambda = 0$, or $\psi_\lambda =$ constant, and

(ii) the fluid is barotropic, $\text{grad}\,\rho \times \text{grad}\,p = \underset{\sim}{0}$ or λ can be considered a function of ρ and/or p only $[\text{grad}\,\lambda = (\partial\lambda/\partial p)\,\text{grad}\,p + (\partial\lambda/\partial\rho)\,\text{grad}\,\rho]$,

then the potential vorticity for each particle remains constant. Comparing (2.13), (2.14) with (2.8) it is now plausible that by appropriately selecting λ, equation (2.13) will generate the special case (2.8).

Consider the equations defining the top and bottom boundaries, $F_S \equiv \zeta(\underset{\sim}{x},t) - z =$ const. and $F_B \equiv H(\underset{\sim}{x}) + z =$ const., relations which must be valid for all times. Hence, F_S, F_B, $F_S + F_B$ and all product combinations of these are candidates for λ, which make the potential vorticity a conserved quantity. In particular

$$\lambda \equiv \frac{F_B}{F_S + F_B} = \frac{z + H}{\zeta + H} \qquad (2.15)$$

is conserved along particle paths. Thus

$$\pi_S = \frac{\underset{\sim}{\omega} + 2\underset{\sim}{\Omega}}{\rho} \cdot \text{grad}\,(\frac{z + H}{\zeta + H}) \qquad (2.16)$$

is also conserved. In a barotropic process where ρ is constant (2.16) is replaced by the *barotropic potential vorticity*

$$\pi_S = (\underset{\sim}{\omega} + 2\underset{\sim}{\Omega}) \cdot \text{grad}\,(\frac{z + H}{\zeta + H}). \qquad (2.17)$$

In the shallow water approximation where the horizontal gradients of H and ζ are small quantities the dominant component of $\text{grad}\,\lambda$ in (2.17) is the z-component. Thus

$$\pi_S = \frac{\omega_z + f}{\zeta + H}, \quad \text{or} \quad \pi_S = \frac{\omega_z + f}{H}, \quad \text{for} \quad \frac{\zeta}{H} \ll 1,$$

satisfy the evolution equations

$$\frac{d}{dt}\left(\frac{\omega_z + f}{\zeta + H}\right) = 0, \quad \text{or} \quad \frac{d}{dt}\left(\frac{\omega_z + f}{H}\right) = 0, \quad \text{for} \quad \frac{\zeta}{H} \ll 1. \tag{2.18}$$

The last equation agress with (2.8).

To derive the evolution equation for TW's from (2.18) consider the mass balance equation[*]

$$\frac{\partial \zeta}{\partial t} + \nabla \cdot (H \underset{\sim}{u}) = 0, \tag{2.19}$$

in which ∇ is the two-dimensional, or horizontal gradient operator. Henceforth this notation will be used throughout, i.e. grad, div, curl are three-dimensional, whereas ∇, $\nabla\cdot$ and $\nabla\times$ are used in two dimensions.

When the *rigid-lid assumption* is made, the first term in (2.19) is ignored. Thus, introducing the mass transport stream function ψ according to

$$\frac{\partial \psi}{\partial y} = -Hu, \quad \frac{\partial \psi}{\partial x} = Hv, \tag{2.20}$$

the continuity equation (2.19) under the rigid lid assumption is satisfied.

In terms of ψ the vertical component of the relative vorticity reads

$$\omega_z = \frac{\partial v}{\partial x} - \frac{\partial u}{\partial y} = \nabla \cdot \frac{1}{H} \nabla \psi.$$

In a two-dimensional barotropic model $(2.18)_2$ then becomes

$$\frac{1}{H}\frac{\partial}{\partial t} \nabla \cdot \left(\frac{\nabla \psi}{H}\right) - \frac{1}{H}\frac{\partial \psi}{\partial y}\frac{\partial}{\partial x}\left(\frac{f}{H}\right) + \frac{1}{H}\frac{\partial \psi}{\partial x}\frac{\partial}{\partial y}\left(\frac{f}{H}\right) = 0. \tag{2.21}$$

Here, non-linear terms have been ignored and f and H have been assumed to be time-independent. Equation (2.21) can be written in the compact vector form

$$\boxed{\frac{\partial}{\partial t}\left(\nabla \cdot \left(\frac{\nabla \psi}{H}\right)\right) + \underset{\sim}{\hat{z}} \cdot \left(\nabla \psi \times \nabla\left(\frac{f}{H}\right)\right) = 0,} \tag{2.22}$$

where $\hat{z}$ is a unit vector in the positive z-direction. Because of its importance we list once more the assumptions on which equation (2.22) is based. They are:

(i)   Processes must be adiabatic and
(ii) barotropic,
(iii) the hydrostatic pressure assumption,
(iv) the shallow water assumption and
(v)   the rigid-lid assumption must hold.

---

[*] *This equation can be obtained from the continuity equation div $\underset{\sim}{u} = 0$ by integrating it from z = -H to z = ζ and assuming that $\underset{\sim}{u}$ does not vary with z. This must strictly be so when (i) the hydrostatic pressure assumption is invoked and (ii) processes are barotropic.*

Variants of (2.22), when some of these conditions are relaxed, will be discussed below.

The boundary condition that must be imposed is the *no-flux condition* through the shore of the basin. This can be expressed as $\partial\psi/\partial s = 0$, where s is the arc length along the shore. Thus

$$\psi = \text{constant, along the shore} \tag{2.23}$$

is the requested boundary condition.

In simply connected domains, or domains with one interior island, the constant in (2.23) is irrelevant and equations (2.22), (2.23) may be written as

$$T[\psi] \equiv \frac{\partial}{\partial t} E[\psi] + J[\psi, \frac{f}{H}] = 0, \quad \text{in } D,$$
$$\psi = 0, \quad \text{on} \quad \partial D, \tag{2.24}$$

where D is a two-dimensional domain and ∂D its boundary, while

$$E[\psi] \equiv \nabla \cdot (\frac{\nabla\psi}{H}),$$

and
$$J[\psi, \frac{f}{H}] \equiv \frac{\partial\psi}{\partial x}\frac{\partial}{\partial y}(\frac{f}{H}) - \frac{\partial\psi}{\partial y}\frac{\partial}{\partial x}(\frac{f}{H}) \tag{2.25}$$

are operators. E[.] is an elliptic and J[.,.] the Jacobian operator.

We now prove the following statement: If J[.,.] is identically zero, (2.24) does not admit wave-like solutions. Indeed with J[.,.] = 0, we may set
$$\psi = \tilde{\psi} \exp(i\omega t)$$

and then obtain from (2.24)

$$E[\tilde{\psi}] = 0, \quad \text{in } D,$$
$$\tilde{\psi} = 0, \quad \text{on} \quad \partial D. \tag{2.26}$$

According to the maximum principle (Protter & Weinberger, 1967, p. 61) any non-constant $\tilde{\psi}$ that obeys (2.26) must assume its maximum and minimum on ∂D which are both zero. Hence (2.26) admits only the solution $\tilde{\psi} = 0$ in D.

The requirement $J[.,.] \neq 0$ implies that in a *rotating frame* one of the following cases is satisfied:

- The case $\nabla f \neq 0$, $\nabla H = 0$ leads to *planetary Rossby waves*. They are significant in global atmospheric and ocean wave dynamics.

- The case $f \neq 0$, $\nabla f = 0$ but $\nabla H \neq 0$ distinguishes *topographic Rossby waves*. In this case the basin must be distant from the equator ($f \neq 0$) and its North-South extent L ought to be small enough that $\beta L = (\partial f/\partial y)L \ll f$. This requires $L \leq 500$ km.

- When $f = 0$, $\nabla f \neq 0$ but $\nabla H = 0$ wave-like solutions are called *equatorial planetary Rossby waves*.

- The case $f \neq 0$, $\nabla f \neq 0$, $\nabla H \neq 0$ characterizes *planetary topographic Rossby waves*. They arise in oceans distant from the equator and describe large scale dynamical events.

- Finally, when $f = 0$, but $\nabla f \neq 0$, $\nabla H \neq 0$ one limits attention to regions close to the equator. These waves may be called *equatorial planetary topographic Rossby waves*.

Steady solutions of the TW-equation (2.24) are described by

$$J\left[\psi, \frac{f}{H}\right] = 0, \quad \text{in } D,$$
$$\psi = 0, \quad \text{on } \partial D. \tag{2.27}$$

Any solution of (2.27) is a field of mass transport currents which is tangential to the isotrophes, i.e. lines $f/H = $ constant.

Indeed, the Jacobian operator can be replaced by

$$\nabla\psi \times \nabla\left(\frac{f}{H}\right) = 0,$$

which means that both gradients are parallel. Since $\nabla\psi$ and $\nabla(f/H)$ represent the orthogonal trajectories of the respective contour lines, the current vectors (orthogonal to $\nabla\psi$) are parallel to the isotrophes (orthogonal to $\nabla(f/H)$).

This analysis also provides the basic understanding of the restoring force mechanism of vorticity waves. Whenever $\psi(\underset{\sim}{x},t)$ is such that $\nabla\psi$ yields a value $J[\psi, f/H] \neq 0$, in other words, whenever the mass transport stream lines are not parallel to the contour lines $f/H = $ constant, the operator $\partial E[\psi]/\partial t$ of the wave equation (2.24) acts to restore this parallelicity. This is why vorticity waves tend to follow the isotrophes.

A last and potentially useful property of the TW-equation (2.24) is obtained, when the equations are made dimensionless. Accordingly, let L_0 and H_0 be typical length and depth scales and ψ_0 a characteristic value for the mass transport stream function. Introducing the transformations

$$\psi = \psi_0 \psi', \quad (x, y, H) = (L_0 x', L_0 y', H_0 h'), \quad t = f^{-1} t',$$

it is straightforward to demonstrate that (2.24) also holds for the dimensionless variables as follows:

$$\frac{\partial}{\partial t'} E'[\psi'] + J'[\psi', 1/h'] = 0, \quad \text{in } D'$$
$$\psi' = 0, \quad \text{on } \partial D'.$$

Because this boundary value problem does not contain any scale dependent terms, i.e. the lengths L_0 and H_0 do not appear, we conclude that solutions to the TW-equation must exhibit a *scale invariance*: Periods and waveforms of the mass transport stream function are the same for all geometrically similar basin profiles, *irrespective of their size*. The same invariance also holds for velocities if these are scaled according to $\psi_0 = L_0 H_0 U_0$, where U_0 is the velocity scale. However, in the individual case the validity of the approximations in deriving (2.24) need be checked.

The scale invariance of the TW-equation is the result of the imposition of the rigid-lid assumption. This can be seen as follows: Consider the shallow water equations

$$\frac{\partial \zeta}{\partial t} + \nabla \cdot (H \underset{\sim}{u}) = 0,$$

$$\frac{\partial \underset{\sim}{u}}{\partial t} + f \, \hat{\underset{\sim}{z}} \times \underset{\sim}{u} + g \nabla \zeta = 0,$$

and eliminate $\underset{\sim}{u}$ to deduce the single equation for the surface elevation

$$\nabla \cdot (H \nabla \zeta_t) + f \, J[H, \zeta] - \frac{1}{g} \, L \, \zeta_t = 0, \qquad (2.28)$$

The first two terms represent the TW-operator for the surface elevation, the last term on the left is due to the deformation of the free surface. Owing to the definition of $L \equiv \partial^2/\partial t^2 + f^2$, (2.28) is a third order partial differential equation in time which admits three wave-type solutions. When $f = 0$ or $J = 0$ the degree of the equation is reduced; in this case the TW-solution is eliminated and the remaining solutions represent gravity waves. Furthermore, in the low-frequency approximation, $L = f^2$, the equation becomes first order in time and only the TW-solution survives.

Writing (2.28) in dimensionless form as demonstrated above we obtain

$$\nabla \cdot (h \nabla \zeta_t) + J[h, \zeta] - \left(\frac{L}{R}\right)^2 L \, \zeta_t = 0 \qquad (2.29)$$

where now all variables and operators are dimensionless, $R^2 = f^2/(g \, H_0)$ is the external Rossby radius, L and H_0 are typical horizontal and vertical length scales[*]. The first two terms on the lhs of (2.29) are

---

[*] *Strictly speaking, H_0 should be the maximum depth relative to the deformed equilibrium surface of the rotating body. With the condition of volume preservation of the water body, L and H_0 are then related. For a paraboloid of latus rectum L and depth D_0 it may be shown that $H_0 = (1 - f^2 L/2g) D_0$, where D_0 is now the rotation independent scale depth. This recognition yields alterations in the definition of the Rossby radius and the frequency relation (3.8) below, see Miles & Ball (1963).*

scale invariant, but obviously the last terms is not, because

$$\frac{L^2}{R^2} = (\frac{L}{H_0})(\frac{f^2 L}{g}),$$

of which the second factor is size dependent. It follows that the scale independence of the TW-equation emerges because the Rossby radius is large; the low-frequency assumption alone as sometimes claimed does not suffice.

2.3 Baroclinic coupling - the two-layer model

a) Prerequisites

Vertical temperature profiles in stratified lakes can be subdivided roughly into three parts (from top to bottom):

Epilimnion: layer with an average surface temperature of about 18°C and several meters depth,

Metalimnion: layer containing the thermocline and experiencing strong temperature gradients,

Hypolimnion: layer with a lower temperature of about 6°C and several tens of meters depth.

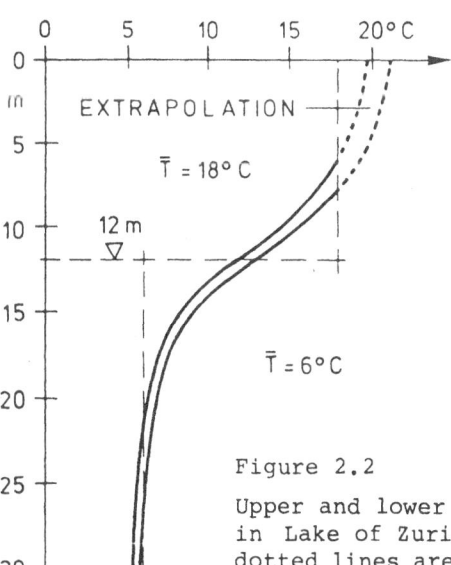

This typical stratification is mainly found during summer periods, when the surface layer is heated by solar irradiation. In a first approximation this situation is simplified by introducing a two-layer system of which the interface represents the position of the thermocline (*Figure 2.2*). Subsequently, the depth of the upper layer will be assumed much smaller than that of the lower layer.

Figure 2.2

Upper and lower bound temperature profiles as measured in Lake of Zurich during August/September 1978. The dotted lines are extrapolations. Also shown are the two layer approximations with density discontinuity at 12 m depth and upper and lower layer temperatures 18°C and 6°C, respectively. [From Hutter, 1984a]

Motions occur in both layers and are subject to a coupling by the thermocline. As we shall show lateron, this coupling mechanism is weak in the sense that it is mainly one-way, i.e. the motion of the thermocline is driven by the barotropic transport. If the velocity fields in the two layers are unidirectional the motion is barotropic, if they are in opposite directions it is baroclinic.

The configuration of the lake and the notation is summarized in *Figure 2.3*. Important in the depicted geometry are the vertical side walls that extend beyond the thermocline well into the hypolimnion. Application must, therefore, be limited to lakes with steep shores.

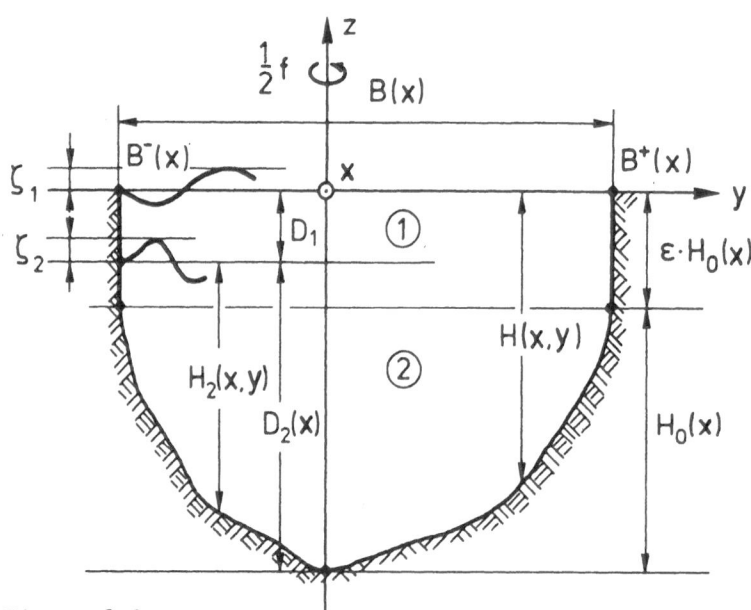

Figure 2.3

Side view of a cross section of the two-layer lake in its natural coordinate system (x,y,z). Upper and lower layer variables are denoted by an index 1 or 2, respectively. The lake is within a rotating system of spatially constant angular velocity $\frac{1}{2}\,f$.

Lake topography varies in space only in the lower layer, i.e. the upper layer is confined by two vertical side walls, which must exceed the depth of the thermocline, so $H(\underset{\sim}{x}) > D_1$ for all $\underset{\sim}{x}$. We accept the varying of the side walls with $\underset{\sim}{x}$ because of analytical simplicity.

b) *Two-layer equations*

Basic idea in obtaining a description of the physical behavior of our system is to formulate equations which describe conservation of mass, momentum and energy for the individual layers. Thermodynamic effects will be neglected in this study. The evolving nonlinear system is linearized by the assumption of small Rossby numbers. Furthermore, surface elevations ζ are thought to be small in comparison to the depth of the upper layer. Turbulence will be ignored but wind stress, distributed over the thin upper layer, and acting as a driving force will be considered. Under these approximations, the equations of motion in components of a Cartesian system take the form (Mysak, 1984, p. 87, Mysak et al., 1985)

$$
\left.
\begin{aligned}
u_{1t} - f v_1 &= -g \zeta_{1x} + \tau^x/(\rho_1 D_1), \\
v_{1t} + f u_1 &= -g \zeta_{1y} + \tau^y/(\rho_1 D_1), \\
D_1 (u_{1x} + v_{1y}) &= \zeta_{2t} - \underline{\zeta_{1t}},
\end{aligned}
\right\}
\tag{2.30}
$$

$$
\left.
\begin{aligned}
u_{2t} - f v_2 &= -g \zeta_{1x} - g'(\zeta_{2x} - \underline{\zeta_{1x}}), \\
v_{2t} + f u_2 &= -g \zeta_{1y} - g'(\zeta_{2y} - \underline{\zeta_{1y}}), \\
(H_2 u_2)_x + (H_2 v_2)_y &= -\zeta_{2t},
\end{aligned}
\right\}
\tag{2.31}
$$

where g' is the reduced gravity $g' = g(\rho_2 - \rho_1)/\rho_2$. Everything that follows can be directly derived from equations (2.30)-(2.31).

c) Approximations

We will now transform the above equations and introduce further approximations which will make it apparent why the conservation law of potential vorticity is still a reasonable approximation for vorticity waves when barotropic-baroclinic coupling is present.

α) Rigid-lid approximation

It is known that to every wave type of the above system there exists an internal and an external variant. The periods of the latter are generally much smaller than those of the former and, by applying the rigid lid approximation, the external modes are impeded. This means, that compared to the interface elevation any surface elevation can be neglected, i.e. the underlined terms in (2.30) and (2.31) are ignored. With this, it follows from the mass conservations $(2.30)_3$ and $(2.31)_3$ that the quasi-solenoidal velocity field can be replaced by the stream function through which the components of the integrated transport are given by

$$
-\psi_y = D_1 u_1 + H_2 u_2, \qquad \psi_x = D_1 v_1 + H_2 v_2.
\tag{2.32}
$$

ψ is called the barotropic or mass transport stream function. Equations (2.30)-(2.31) can be transformed into a compact system in the variables ψ and $\zeta_2 \equiv \zeta$. Assuming a constant Coriolis parameter f the result reads.

$$\nabla \cdot (H^{-1} \nabla \psi_t) + f(\nabla \psi \times \nabla H^{-1}) \cdot \hat{\underset{\sim}{z}} = -g' D_1 (\nabla \zeta \times \nabla H^{-1}) \cdot \hat{\underset{\sim}{z}} \tag{2.33}$$

$$+ \frac{1}{\rho_1} \left[\nabla \times (\underset{\sim}{\tau} H^{-1}) + \frac{H}{D_1} \underset{\sim}{\tau} \times \nabla H^{-1} \right] \cdot \hat{\underset{\sim}{z}} ,$$

$$H \nabla^2 \zeta_t - \frac{H^2}{g' D_1 H_2} L \zeta_t + \frac{D_1}{H_2} \nabla \zeta_t \cdot \nabla H - \frac{f D_1}{H_2} (\nabla \zeta \times \nabla H) \cdot \hat{\underset{\sim}{z}}$$

$$= \frac{1}{g' H_2} \left[\nabla (L \psi) \times \nabla H \right] \cdot \hat{\underset{\sim}{z}} \tag{2.34}$$

$$- \frac{H}{\rho_1 g' D_1} f (\nabla \times L \underset{\sim}{\tau}) \cdot \hat{\underset{\sim}{z}} ,$$

where the operator $L = \partial_{tt} + f^2$ has been introduced.

Mysak et al. (1985) give a detailed discussion of the physics of equations (2.33) and (2.34), which is now briefly summarized. In the absence of stratification (g' = 0) and wind forcing ($\tau = 0$), equation (2.33) reduces to the conservation law of potential vorticity, (2.24). Wind is the external force; the second term on the rhs of (2.33) may therefore be interpreted as the supply of potential vorticity due to wind action. Stratification (g' $\neq$ 0) in a basin with variable topography ($\nabla H \neq 0$) couples the barotropic part of (2.33), namely its lhs, with the baroclinic processes. The first term on the rhs of (2.33) is therefore the production of potential vorticity due to baroclinicity; it represents the influence of the baroclinic effects on the barotropic motion.

By the same argument, equation (2.34) describes the influence of the barotropic processes (terms involving ψ) and the wind (terms involving τ) on the baroclinic motion. Ignoring these barotropic terms results in an equation describing internal waves with a phase speed

$$c_{int}^2 = g' D_1 H_2/H .$$

When $\nabla H = 0$ the third and fourth term on the lhs vanish, and the equation describes classical internal Kelvin and Poincaré waves.

Thus, equations (2.33). and (2.34) exhibit in general a two-way coupling, a baroclinic-barotropic coupling and a barotropic-baroclinic coupling the strengths of which must be estimated by a scale-analysis.

When deriving (2.33) and (2.34) from (2.30)-(2.31) the layer velocities can be expressed in terms of ψ and ζ. The expressions are

$$L \underset{\sim}{u}_1 = \frac{1}{H} \left[\hat{\underset{\sim}{z}} \times \nabla(L\psi) + H_2 g'(\nabla \zeta_t - f \hat{\underset{\sim}{z}} \times \nabla \zeta) + \frac{H_2}{\rho_1 D_1}(\underset{\sim}{\tau}_t - f \hat{\underset{\sim}{z}} \times \underset{\sim}{\tau}) \right],$$

$$L \underset{\sim}{u}_2 = \frac{1}{H} \left[\hat{\underset{\sim}{z}} \times \nabla(L\psi) - D_1 g'(\nabla \zeta_t - f \hat{\underset{\sim}{z}} \times \nabla \zeta) - \frac{1}{\rho_1}(\underset{\sim}{\tau}_t - f \hat{\underset{\sim}{z}} \times \underset{\sim}{\tau}) \right],$$

$$(2.35)$$

which are additively composed of three parts, i.e. a barotropic, a baroclinic and a wind force component. The first are the same (and unidirectional) in both layers, and the second are in opposite directions and
add up to vanishing total transport, reminiscent of barotropic and baroclinic behavior, respectively.

β) Low-frequency approximation

In equations (2.34) ζ appears with a third order time derivative. This
means that (2.34) can contain three types of waves. In fact a more precise analysis shows that there are two (internal) gravity waves and one
topographic wave of which the latter has the longest period. Because we
want to study here topographic waves, we will search for solutions of
(2.33), (2.34) with low frequency ω. For $\omega \ll f$ we may therefore neglect
ω in comparison to f. Thus L reduces to $L = f^2$. Such an approximation,
however, requires that periods are substantially greater than about 17
hours (the period corresponding to f at 45° latitude).

Parenthetically, we might also mention that this approximation holds
only for lakes in which the internal seiche period (of a gravity or Kelvin wave) is considerably smaller than the period of topographic waves.
Since the former increases with the lake dimension and the latter is
size-invariant, the frequencies of gravity waves in larger lakes become
of comparable order to those of topographic waves. For the Lakes of Zurich and Lugano the approximation is appropriate, for Lake Geneva or
larger lakes it may be dubious, see *Table 2.1*.

Lake	Surface length [km]	Period of internal gravity waves [h]	Period of topographic waves [h]	
Lugano	17.2[1]	≤ 28[1]	74[2]	1) Hutter, 1983
Zurich	28[1]	≤ 45[1]	100[2]	2) Mysak, 1985
Geneva	72[3]	≤ 78[4]	$72 - 96$[2]	3) Graf, 1983
				4) Bäuerle, 1985

Table 2.1   The gap between the eigenperiods of internal gravity
and topographic waves depends on the lake dimension.

The situation is nevertheless not as limiting as this statement might
let us surmise, because we shall prove below that for many situations
the baroclinic-barotropic coupling term on the rhs of (2.33) may safely
be ignored. In this case, the TW-equation (2.33) uncouples from (2.34).
Since also boundary conditions will be shown to be free of this baroclinic-barotropic coupling, solutions to the TW-problem can be obtained
without solving the inertial gravity wave problem. The assumption $|\omega| \ll f$
need not necessarily be invoked.

d) Scale analysis

Information about the orders of magnitude of the various coupling terms
in (2.33) and (2.34) is obtained by constructing dimensionless counter-
parts of these equations via the introduction of scales.

α) Wind forcing

The external forcing mechanism in equations (2.33) and (2.34) is the
wind. To estimate its relative importance consider the identity

$$\nabla \times (\underset{\sim}{\tau} H^{-1}) + \frac{H}{D_1} \underset{\sim}{\tau} \times \nabla H^{-1} \equiv H^{-1}(\nabla \times \underset{\sim}{\tau}) + (\nabla H^{-1}) \times \underset{\sim}{\tau} + \frac{H}{D_1} \underset{\sim}{\tau} \times \nabla H^{-1}. \qquad (2.36)$$

The first term on the right can be neglected in comparison to the others,
because the atmospheric length scale is in general much larger than the
lake dimensions. Such a statement is tantamount to ignoring spatial variations of wind stress over the lake's domain. Further, comparing the
last two terms it is seen that they differ by an order D_1/H which, in view
of our basic assumption, is small (cf. Table 2.2). Consequently only the
last term of (2.36) survives. In a way this is a strange result: As far
as the barotropic contribution of the motion is concerned, only a lake
with variable topography can be affected by the wind. This leads to the
conclusion that the assumption on atmospheric length scales may be doubt-
ful. Indeed, a varying topography in the vicinity of the lake may play
a significant role as it can modify regional winds with atmospheric
length scales to local winds with smaller length scales. An example is
the topography around Lake of Lugano; but experimental evidence for the
wind stress curl to be significant is so far lacking.

β) Gratton's scaling

Gratton (1983) and Gratton & Le Blond (1986) consider lake stratifica-
tions with $D_1 \ll D_2$, i.e. a thin upper layer lies on the top of a deep

hypolimnion. For this case they found that the baroclinic effect on the barotropic motion is of order D_1/D_2 smaller than the barotropic effect on the baroclinic motion. So, to order D_1/D_2 the coupling only arises as a forcing of the baroclinic motion by the barotropic mass transport.

Before we demonstrate this result let us point out its significance. The one-way coupling means that traces of the topographically induced motion can be observed by measuring baroclinic quantities such as temperature-time series of thermistor chains, moored within the metalimnion, The description of the observations in Lakes of Lugano and Zurich in Chapter 1 are based on such temperature-time series.

We now introduce the following set of nondimensional variables:

$$\psi: = \psi_0 \, \psi', \quad \zeta: = \zeta_0 \, \zeta', \quad \underset{\sim}{\tau}: = \tau_0 \, \underset{\sim}{\tau}',$$

$$(x,y): = L(x',y'), \quad t: = f^{-1} \cdot t', \tag{2.37}$$

$$H: = Dh' = (D_1 + D_2)h', \quad H_2: = D_2 \, h_2,$$

where the primed variables are non-dimensional; L is a typical length scale of the considered waves (e.g. half the lake length). Higher wave modes, where cross variations are important, may require a (x,y)-scaling which is different for each spatial direction, but this will not be considered. With (2.37) we obtain the scaled equations

$$\nabla \cdot (h^{-1} \, \nabla\psi_t) + (\nabla\psi \times \nabla h^{-1}) \cdot \hat{\underset{\sim}{z}} = -C_1(\nabla\zeta \times \nabla h^{-1}) \cdot \hat{\underset{\sim}{z}} + \left(\frac{L \, D\tau_0}{f\rho_1 \, D_1 \, \psi_0}\right) (h\underset{\sim}{\tau} \times \nabla h^{-1}) \cdot \hat{\underset{\sim}{z}}, \tag{2.38}$$

$$\frac{1}{h}(\nabla^2 \zeta_t - \frac{L^2 h}{R_i^2 \, h_2} L\zeta_t) - \frac{D_1}{D_2 \, h_2} \nabla\zeta_t \cdot \nabla h^{-1} + \frac{D_1}{D_2 \, h_2} (\nabla\zeta \times \nabla h^{-1}) \cdot \hat{\underset{\sim}{z}} = -C_2 \, h_2^{-1} (\nabla\psi \times \nabla h^{-1}) \cdot \hat{\underset{\sim}{z}}$$
$$- \frac{\tau_0 \, L}{\rho_1 \, g' \, D_1 \, \zeta_0} \frac{1}{h} (\nabla \times L \, \underset{\sim}{\tau}) \cdot \hat{\underset{\sim}{z}}, \tag{2.39}$$

where now $L = \partial_{tt} + 1$ and the coupling coefficients are given by

$$C_1 = \frac{g' \, D_1 \, \zeta_0}{f\psi_0}, \quad C_2 = \frac{f\psi_0}{g' \, D_2 \, \zeta_0} = \frac{D_1}{D_2} \cdot \frac{1}{C_1}, \tag{2.40}$$

and $R_i = (g' D_1 D_2/Df^2)^{1/2}$ is the internal Rossby radius. Note that in (2.38) and (2.39) we have dropped the primes on the scaled (nondimensional) variables.

Let us now suppose that (2.38) and (2.39) are strongly coupled, i.e. that C_1 and C_2 are both $O(1)$. Then (2.40) implies that

$$\zeta_0 = O\left(\frac{f\psi_0}{g' \, D_1}\right) \quad \text{and} \quad \zeta_0 = O\left(\frac{f\psi_0}{g' \, D_2}\right). \tag{2.41 \& 2.42}$$

We observe that independent of the ψ_0-scale, (2.41) and (2.42) are consistent only if $D_1/D_2 = O(1)$. Since we are concerned with the case $D_1 \ll D_2$, it follows that C_1 and C_2 cannot both be of order unity, i.e. that (2.38) and (2.39) are only weakly coupled. Suppose we assume that (2.41) applies and thus choose

$$\frac{\zeta_0}{\psi_0} = \frac{f}{g' \, D_1} \tag{2.43}$$

as the scaling for the ratio ζ_0/ψ_0. Then $C_1 = 1$ and $C_2 = D_1/D_2 \ll 1$. Therefore to $O(D_1/D_2)$, equation (2.39) reduces to

$$\left(\nabla^2 - \left(\frac{L}{R_i}\right)^2 L\right) \zeta_t = -\frac{\tau_0 \, L}{\rho_1 \, g' \, D_1 \, \zeta_0} \, (\nabla \times L\underset{\sim}{\tau}) \cdot \underset{\sim}{\hat{z}} , \tag{2.44}$$

where we have used $h/h_2 = 1 + O(D_1/D_2)$. (2.44) is a wave equation forced by the wind stress curl, but the scale choice (2.43) leads to an unrealistically large value for the ζ_0 scale (about $\zeta_0 \geq 50$ m, which is several times the upper layer depth for most lakes[*]).

Hence we are compelled to choose the scaling (2.42) (Gratton's choice, which was based on data from the Strait of Georgia, British Columbia). Putting

$$\frac{\zeta_0}{\psi_0} = \frac{f}{g' \, D_2} , \tag{2.45}$$

we find $C_2 = 1$ and $C_1 = D_1/D_2 \ll 1$. We choose the ζ_0 scale by setting the coefficient of the wind stress term in (2.38) equal to unity, which gives

$$\psi_0 = \frac{L D \tau_0}{f \, \rho_1 \, D_1} . \tag{2.46}$$

Substituting (2.46) into (2.45) gives the scale ζ_0 in terms of the wind stress scale τ_0:

$$\zeta_0 = \frac{L D \tau_0}{\rho_1 \, g' \, D_1 \, D_2} = \frac{\psi_0 f}{g' \, D_2} , \tag{2.47}$$

which yields a realistic order of magnitude[**]. Using (2.46) and (2.47) in (2.38) and (2.39), we obtain, correct to $O(D_1/D_2)$

$$\nabla \cdot (h^{-1} \nabla \psi_t) + (\nabla \psi \times \nabla h^{-1}) \cdot \underset{\sim}{\hat{z}} = (h\underset{\sim}{\tau} \times \nabla h^{-1}) \cdot \underset{\sim}{\hat{z}} , \tag{2.48}$$

$$(\nabla^2 - S^{-1} L) \zeta_t = -(\nabla \psi \times \nabla h^{-1}) \cdot \underset{\sim}{\hat{z}} - (\nabla \times L\underset{\sim}{\tau}) \cdot \underset{\sim}{\hat{z}} , \tag{2.49}$$

---

[*] *With $f = 10^{-4}$ s^{-1}, $g' = 0.02$ m s^{-2}, $D_1 = 10$ m, $D_2 = 270$ m and $\psi_0 = U \cdot L \cdot (D_1 + D_2) = 0.03 \cdot 10^4 \times 270$ m^3 s^{-1}, where U is a velocity scale (approximately 3 cm s^{-1} for Lake of Lugano) and $L = 10^4$ m, one obtains $\zeta_0 = 40$ m.*

[**] *With $f = 10^{-4}$ s^{-1}, $g' = 0.02$ m s^{-2}, $\psi_0 = 7 \times 10^4$ m^3 s^{-1} and $D_2 = 270$ m, (2.47) yields $\zeta_0 = 1$ m. Alternatively, using $\tau_0 = \rho_{air} \, c_d \, U_w^2$ with $\rho_{air} = 1.29$ kg m^{-3}, $c_d = 1.85 \times 10^{-3}$ (an average value for lakes during summer, see Simons, 1980, p. 92), and $U_w = 4$ m s^{-1}, we find $\tau_0 = 0.038$ Nm^{-2} and hence according to (2.47), $\zeta_0 = 1.5$ m and according to (2.46), $U = \tau_0/(\rho_1 \, f D_1) = 3.8$ cm s^{-1}. Both values are typical observations in the Lake of Zurich and Lake of Lugano, see Table 2.2.*

as the appropriate non-dimensional equations for ψ and ζ. In (2.49) note that we have introduced the stratification parameter S, defined as

$$S = (\frac{R_i}{L})^2. \tag{2.50}$$

For the derivation of (2.49) it is important that $h_2 \neq 0$ (as illustrated in *Figure 2.3*). If $h_2 = 0$, the third and fourth terms on the left side of (2.39) are not uniformly $O(D_1/D_2)$ and hence could not be neglected. If an elliptic paraboloid contained a two-layer fluid, then clearly $h_2 = 0$ where the interface touches the basin wall and (2.49) would not be valid. Thus Ball's (1965) solution for second-class waves in an elliptic paraboloid could not be easily extended to the stratified case by our analysis.

The derivation of (2.48) and (2.49) follows Mysak et al. (1985) but is more general in that the low frequency assumption has not been invoked and the wind stress curl has not been ignored. With these two additional assumptions l would be replaced by $l = 1$ and the last term in (2.49) would be missing. As stated above these assumptions are not needed to achieve the decoupling of the barotropic motion from the baroclinic influence.

Substituting (2.37) and the scaling (2.42) into (2.35) and using the scale $\psi_0 = ULD$, as before, we obtain the following formulae for the velocities:

$$l\underset{\sim}{u}_1 = \frac{1}{h}\left[\hat{\underset{\sim}{z}} \times \nabla L\psi + h_2\left((\nabla\zeta_t - \hat{\underset{\sim}{z}} \times \nabla\zeta) + \frac{D_2}{D}(\underset{\sim}{\tau}_t - \hat{\underset{\sim}{z}} \times \underset{\sim}{\tau})\right)\right], \tag{2.51}$$

$$l\underset{\sim}{u}_2 = \frac{1}{h}\left[\hat{\underset{\sim}{z}} \times \nabla L\psi - \frac{D_1}{D_2}\left((\nabla\zeta_t - \hat{\underset{\sim}{z}} \times \nabla\zeta) + \frac{D_2}{D}(\underset{\sim}{\tau}_t - \hat{\underset{\sim}{z}} \times \underset{\sim}{\tau})\right)\right]. \tag{2.52}$$

To $O(D_1/D_2)$ these can be approximated by

$$l\underset{\sim}{u}_1 = \frac{1}{h}(\hat{\underset{\sim}{z}} \times \nabla L\psi + h(\nabla\zeta_t - \hat{\underset{\sim}{z}} \times \nabla\zeta + \underset{\sim}{\tau}_t - \hat{\underset{\sim}{z}} \times \underset{\sim}{\tau})), \tag{2.53}$$

$$l\underset{\sim}{u}_2 = \frac{1}{h}\hat{\underset{\sim}{z}} \times \nabla\psi. \tag{2.54}$$

Thus for deep lakes, the lower layer current associated with topographic waves is essentially barotropic, whereas the upper layer current consists of a barotropic part, a baroclinic part and a contribution directly forced by the wind. Hence we conclude that the current motions are generally surface intensified.

In *Table 2.2* we collect some data pertinent to the above estimates. Values are given for the layer thickness and density difference of the summer stratification for three Swiss lakes from which Rossby radii, stratification parameters and typical values of the thermocline elevation can be computed. Accordingly, neglecting $O(D_1/D_2)$ terms is certainly justified

Lake	D_1	D_2^{mean}	D_2^{max}	$\dfrac{D_1}{D_2^{mean}}$	$\dfrac{\rho_2-\rho_1}{\rho_2}$	R_i	half length	S^{-1}	ζ_0
	[m]	[m]	[m]			[km]	[km]		[m]
Lugano	$10^{1)}$	$183^{1)}$	$278^{2)}$	0.055	$1.91 \cdot 10^{-3\ 1)}$	4.05	8.6	4.5	$1.8^{5)}$
Zurich	$12^{1)}$	$52^{1)}$	$124^{2)}$	0.231	$1.75 \cdot 10^{-3\ 1)}$	4.13	14	11.5	$2.9^{5)}$
Geneva	$15^{4)}$	$153^{4)}$	$310^{3)}$	0.098	$1.41 \cdot 10^{-3\ 4)}$	4.24	36	72.1	$6.9^{5)}$

1) *Hutter, 1984c, p. 78*   4) *Bäuerle, 1984*
2) *Hutter, 1983, p. 108*   5) *Computed using (2.47)*
3) *Graf, 1983, p. 64*

Table 2.2   Properties of some Swiss Lakes.

for Lake of Lugano and still reasonable for the other lakes. Moreover, the thermocline-elevation amplitude ζ_0 is smaller than D_1 in all three cases, a fact which gives some confidence in the scaling procedure.

e) *Boundary conditions*

To solve (2.48) and (2.49) in some domain $\mathcal{D}$ for a given $\underset{\sim}{\tau}$, we have to prescribe initial values for ψ and ζ and the boundary conditions on $\partial\mathcal{D}$. The first boundary condition we impose is that the total mass flux normal to $\partial\mathcal{D}$ must vanish: in non-dimensional variables this can be written as $\hat{n} \cdot (D_1 \underset{\sim}{u}_1 + D_2 h_2 \underset{\sim}{u}_2) D^{-1} = 0$ on $\partial\mathcal{D}$, where $\hat{n}$ is a unit vector perpendicular to $\partial\mathcal{D}$. On substituting for $\underset{\sim}{u}_1$ and $\underset{\sim}{u}_2$ from (2.51) and (2.52), this reduces to[*]

$$\hat{n} \cdot (\hat{\underset{\sim}{z}} \times \nabla\psi) = 0, \qquad \text{on} \quad \partial\mathcal{D}. \qquad (2.55)$$

Since $\hat{n} \cdot (\hat{\underset{\sim}{z}} \times \nabla\psi) = (\hat{n} \times \hat{\underset{\sim}{z}}) \cdot \nabla\psi = \hat{\underset{\sim}{s}} \cdot \nabla\psi$, where $\hat{\underset{\sim}{s}}$ is a unit vector tangential to $\partial\mathcal{D}$, (2.55) implies $\partial\psi/\partial s = 0$ on $\partial\mathcal{D}$ and hence $\psi = \text{constant}$ on $\partial\mathcal{D}$. Thus in a simply connected domain, without loss of generality, we take

$$\psi = 0, \qquad \text{on} \quad \partial\mathcal{D}. \qquad (2.56)$$

Next we require $\hat{\underset{\sim}{n}} \cdot \underset{\sim}{u}_i = 0$ on $\partial\mathcal{D}$ for each layer i. Upon again using (2.51) and (2.52), together with (2.55), we find

$$\frac{\partial}{\partial n} \frac{\partial\zeta}{\partial t} - \frac{\partial\zeta}{\partial s} = - \frac{D_2}{D} (\hat{n} \cdot \underset{\sim}{\tau}_t - \hat{\underset{\sim}{s}} \cdot \underset{\sim}{\tau}) = -\hat{n} \cdot \underset{\sim}{\tau}_t + \hat{\underset{\sim}{s}} \cdot \underset{\sim}{\tau}, \qquad \text{on} \quad \partial\mathcal{D}, \qquad (2.57)$$

to $O(D_1/D_2)$.

---

[*] *These equations actually imply a statement regarding $L \cdot (D_1 \underset{\sim}{u}_1 + D_2 h_2 \underset{\sim}{u}_2) \cdot \hat{n}$ rather than the mass transport itself. However if $Lm = 0$ along $\partial\mathcal{D}$ for all time, then necessarily $m = 0$ as well.*

The boundary condition (2.55) applies, whether the simplifying assumptions $D_1 \ll D_2$ and $\omega^2 \ll f^2$ are imposed or not. Because (2.48) supposes D_1 to be small in comparison to D_2 we conclude that the barotropic part of the motion can be determined without simultaneously also determining the baroclinic response. However, if the corresponding barotropically driven baroclinic currents or thermocline elevations are to be determined, then equation (2.49) subject to the boundary condition (2.57) must also be solved. Since (2.49) is a forced wave equation, this by itself is a formidable problem. For *weak stratification* (S small) simplifications are possible. This is the case for most Swiss lakes (compare *Table 2.2*).

To introduce this additional simplification we note that our scales have been chosen such that dimensionless gradients are order unity. Hence, we expect ∇^2 to be $O(1)$ whereas S^{-1} is large. On the lhs of equation (2.49) we may thus ignore ∇^2 in comparison to $S^{-1} L\, h/h_2$, implying

$$\zeta_t = S\{L^{-1} (\hat{\underset{\sim}{z}} \times \nabla\psi) \cdot \nabla h^{-1} + (\nabla \times \underset{\sim}{\tau}) \cdot \hat{\underset{\sim}{z}}\}, \qquad (2.58)$$

where L^{-1} is the inverse operator of L. Equation (2.58) can be described as the *geometric optics approximation for* ζ. Along the shore ∂D we may assume a constant depth; then ∇h^{-1} is parallel to $\hat{\underset{\sim}{n}}$, the unit normal vector along ∂D, and the first term in the curly bracket vanishes[*]. With non-vanishing wind stress the emerging equation is not consistent with (2.57). For the *unforced* problem, however, (2.58) implies

$$\zeta(\underset{\sim}{x},t) = 0, \qquad \text{along} \qquad \partial D,$$

which is consistent with (2.57) provided that the term $\partial^2 \zeta/\partial n\, \partial t$ is ignored. This omission is justified in the low-frequency approximation.

We conclude: the geometric optics approximation is only consistent in the low-frequency limit. In all other cases the baroclinic coupling should be computed with the full equations (2.48),(2.49) and (2.57).

2.4 Continuous stratification

a) *Modal equations*

Most lakes in the temperate belt are strongly stratified during the warm period of the year. This has already been outlined in section 2.3a. The two-layer model can only nearly approximate the internal dynamics of a lake that permits a clear distinction of an epi- and a hypolimnion. It is

---

[*] *Recall that* $(\hat{\underset{\sim}{z}} \times \nabla\psi) \cdot \nabla h^{-1} = 0$ *for all times implies that* $L^{-1}\{(\hat{\underset{\sim}{z}} \times \nabla\psi) \cdot \nabla h^{-1}\} = 0$.

important to investigate to which extent inferences from the pure barotropic or two-layer model of section 2.3 can be carried over to a lake
with continuous stratification. The oceanographic literature is rich in
studies of low-frequency processes in a stratified medium. A general
result, common to all studies, is that increasing the stratification increases the frequencies of the considered long-period motion. This has
been shown by Wang & Mooers (1976), Clarke (1977) and Huthnance (1978).
A review can be found in Mysak (1980a).

Here in this section our aim is to analyse, how and how strongly the
individual baroclinic modes are coupled among each other and how the
baroclinic part of the motion couples with the barotropic processes. To
this end, consider the vertically integrated balance laws of mass and
momentum (Hutter, 1984a, p. 20-21)

$$\frac{\partial \zeta}{\partial t} + \nabla \cdot \underset{\sim}{V} = 0,$$

$$\frac{\partial \underset{\sim}{V}}{\partial t} + f \, \hat{\underset{\sim}{k}} \times \underset{\sim}{V} = - \frac{H+\zeta}{\rho^\star} \, \nabla p_e + \underset{\sim}{F},$$

(2.59)

in which ζ is the surface elevation, $\underset{\sim}{V}$ the transport, $\hat{\underset{\sim}{k}}$ a unit vector
pointing upwards, $\rho^\star$ a constant reference density and

$$p_e = p^{atm} + \rho^\star g(\zeta - z),$$

$$p_i = \rho^\star g \int_z^\zeta \sigma(z') \, dz' + p', \qquad \sigma \equiv \frac{\rho_0 - \rho^\star}{\rho^\star},$$

(2.60)

$$\underset{\sim}{F} = - \frac{1}{\rho^\star} \int_{-H}^\zeta \nabla p_i \, dz - \nabla \cdot \int_{-H}^\zeta (\underset{\sim}{v} \otimes \underset{\sim}{v} + \underset{\sim}{\Gamma}) \, dz + \frac{\underset{\sim}{\tau}_{wind} - \underset{\sim}{\tau}_{bottom}}{\rho^\star}.$$

Here, p_e is the external and p_i the internal pressure. The latter consists of the dynamic baroclinic pressure p' and the quasistatic baroclinic pressure due to the density anomaly $\sigma(z)$, which is referred to
a reference profile of density $\rho_0(z)$ of a *stably stratified state* at rest.
The hydrostatic pressure assumption has not been made in equations (2.60);
this is important. The force $\underset{\sim}{F}$ consists of a contribution of the baroclinic pressure p_i, a term involving advection $(\underset{\sim}{v} \otimes \underset{\sim}{v})$ and turbulent
diffusion $(\underset{\sim}{\Gamma})$, the wind stress and the bottom shear stress. In ensuing
developments we ignore turbulent diffusion $(\underset{\sim}{\Gamma} \simeq 0)$ and advection $(\underset{\sim}{v} \otimes \underset{\sim}{v} \simeq 0)$,
omit bottom shear $(\underset{\sim}{\tau}_{bottom} \simeq 0)$ and spatial variations of the atmospheric
pressure $(\nabla p^{atm} \simeq \underset{\sim}{0})$ and neglect non-linear terms such as $\zeta \nabla \zeta$. If we also
invoke the rigid-lid assumption and thus eliminate the surface gravity
waves, equations (2.59) and (2.60) reduce to

$$\nabla \cdot (H \bar{\underset{\sim}{v}}) = 0,$$

$$\frac{\partial \bar{\underset{\sim}{v}}}{\partial t} + f \, \hat{\underset{\sim}{k}} \times \bar{\underset{\sim}{v}} = - g \nabla \zeta - \frac{1}{\rho * H} \int_{-H}^{0} \nabla p' \, dz + \frac{\underset{\sim}{\tau}}{\rho * H} \tag{2.61}$$

where $H \bar{\underset{\sim}{v}} = \underset{\sim}{V}$ has been introduced and where $\underset{\sim}{\tau} = \underset{\sim}{\tau}_{wind}$. In the next step we introduce the mass transport stream function ψ by setting

$$H \bar{u} = - \psi_y, \qquad H \bar{v} = \psi_x \tag{2.62}$$

and eliminate $\nabla \zeta$ from $(2.61)_2$ by taking the curl of this equation. This transforms (2.61) to the single equation

$$\nabla \cdot (\frac{1}{H} \nabla \psi_t) + J(\psi, \frac{f}{H}) = - \frac{\partial}{\partial x} \left[\frac{1}{\rho * H} \int_{-H}^{0} \frac{\partial p'}{\partial y} \, dz \right] +$$

$$+ \frac{\partial}{\partial y} \left[\frac{1}{\rho * H} \int_{-H}^{0} \frac{\partial p'}{\partial x} \, dz \right] + \nabla \times (\frac{\underset{\sim}{\tau}}{\rho * H}). \tag{2.63}$$

This equation is analogous to (2.33). Notice that it involves no term due to the stratification of the state defined by $\rho_0(z)$. Baroclinic effects are all contained in the dynamic pressure p', but the terms involving p' may also describe deviations from the hydrostatic pressure distribution. We shall interpret p' as being due to baroclinic effects. Thus, in the absence of stratification and without wind forcing, equation (2.63) reduces to the conservation law of potential vorticity, (2.22). Stratification and external winds (the terms of the rhs of (2.63)) act as supplies of potential vorticity. Thus, the first term on the rhs couples the barotropic part of (2.63) to the baroclinic processes. Our experience with the two layer model suggests that this baroclinic coupling is small and can be ignored to lowest order. If this is correct, the barotropic motion can be fully determined from equations $(2.61)-(2.63)$ by simply omitting the p'-dependent terms.

To complete the formulation we still need a system of equations that describes the baroclinic processes and is coupled to the barotropic motion. To deduce it let us consider the Boussinesq approximated adiabatic equations of motion (Hutter, 1984a, equations (3.17), p. 17)

$$u_t - fv = - \frac{1}{\rho *} p'_x,$$

$$\rho'_t + \frac{d\rho_0}{dz} w = 0,$$

$$v_t + fu = - \frac{1}{\rho *} p'_y,$$

$$w_t + \frac{g}{\rho *} \rho' + \frac{1}{\rho *} p'_z = 0. \tag{2.64}$$

$$u_x + v_y + w_z = 0,$$

The first two of these are the horizontal momentum equations, the third expresses continuity, the fourth derives from the adiabatic heat equation, $dT(\rho)/dt = 0$. Finally, the last equation is the vertical momentum equation; if w_t were ignored, the hydrostatic pressure assumption would result from it.

The two equation sets (2.63), describing the transport, and (2.64), representing the vertical details, suggest to split the velocity field into two parts

$$(u, v) = (\bar{u}, \bar{v}) + (\tilde{u}, \tilde{v}), \qquad (2.65a)$$

such that the total transport of (u, v) is incorporated in the barotropic field and hence

$$\int_{-H}^{0} \tilde{u} \, dz = \int_{-H}^{0} \tilde{v} \, dz = 0. \qquad (2.65b)$$

Substitution of (2.65a) in (2.64) and use of (2.61) yields the new set of equations

$$\tilde{u}_t - f\tilde{v} + \frac{1}{\rho*} p'_x - \frac{1}{\rho* H} \int_{-H}^{0} p'_x \, dz = g\zeta_x - \frac{\tau_1}{\rho* H} ,$$

$$\tilde{v}_t + f\tilde{u} + \frac{1}{\rho*} p'_y - \frac{1}{\rho* H} \int_{-H}^{0} p'_y \, dz = g\zeta_y - \frac{\tau_2}{\rho* H} ,$$

$$\tilde{u}_x + \tilde{v}_y + w_z = \frac{H_x}{H} \bar{u} + \frac{H_y}{H} \bar{v}, \qquad (2.66)$$

$$\rho'_t + \frac{d\rho_0}{dz} w = 0,$$

$$w_t + \frac{\rho'}{\rho*} g + \frac{1}{\rho*} p'_z = 0,$$

in which τ_1 and τ_2 are the x- and y-components of the wind stress. These have been written, so that the external wind forcing and the barotropic contributions of the motion appear on the right hand sides of the equations.

Let us now introduce the following expansions of the baroclinic fields:

$$\tilde{u}(x,y,z,t) = \sum_{n=1}^{N} U_n(x,y,t) \, \phi_n\left(\frac{z}{H}\right),$$

$$\tilde{v}(x,y,z,t) = \sum_{n=1}^{N} V_n(x,y,t) \, \phi_n\left(\frac{z}{H}\right),$$

$$w(x,y,z,t) = \sum_{n=1}^{N} W_n(x,y,t) \, \Xi_n\left(\frac{z}{H}\right), \qquad (2.67)$$

$$p'(x,y,z,t) = \sum_{n=1}^{N} P_n(x,y,t) \, \psi_n\left(\frac{z}{H}\right),$$

$$\rho'(x,y,z,t) = \sum_{n=1}^{N} R_n(x,y,t) \, \chi_n\left(\frac{z}{H}\right).$$

Here, $\{\phi_n, \Xi_n, \psi_n, \chi_n\}$, $n = 1, 2, 3, \ldots, N$ are known and their determination will be explained below. To find the evolution equations for U_n, V_n, W_n, P_n and R_n the *Principle of Weighted Residuals* is used. It essentially amounts to evaluating the integrals

$$\int_{-H}^{0} \delta\phi_m^M (2.66\,a,b)\ dz, \quad \int \delta\phi_m^C (2.66\,c)\ dz, \quad \int \delta\phi_m^V (2.66\,d,e)\ dz$$

for arbitrary weighting functions $\delta\phi_m^L (L = M, C, V)$. Inserting (2.67) into these expressions, equations (2.66) can be reduced to the following spatially two-dimensional system of differential equations for the coefficient functions U_n, V_n, W_n, P_n and R_n (for details of the derivation see Appendix A.).

$$A_{mn}^M \left(\frac{\partial U_n}{\partial t} - f\,V_n \right) - \frac{1}{\rho^*} \frac{\partial H/\partial x}{H} B_{mn}\,P_n + \frac{C_{mn}}{\rho^*} \frac{\partial P_n}{\partial x} = \left(g \frac{\partial \zeta}{\partial x} - \frac{\tau_1}{\rho^* H} \right) D_m^M,$$

$$A_{mn}^M \left(\frac{\partial V_n}{\partial t} + f\,U_n \right) - \frac{1}{\rho^*} \frac{\partial H/\partial y}{H} B_{mn}\,P_n + \frac{C_{mn}}{\rho^*} \frac{\partial P_n}{\partial y} = \left(g \frac{\partial \zeta}{\partial y} - \frac{\tau_2}{\rho^* H} \right) D_m^M,$$

$$A_{mn}^C \left(\frac{\partial U_n}{\partial x} + \frac{\partial V_n}{\partial y} \right) - K_{mn} \left(\frac{\partial H/\partial x}{H} U_n + \frac{\partial H/\partial y}{H} V_n \right) - L_{mn} \frac{W_n}{H}$$

$$\qquad = \left(\bar{u} \frac{\partial H/\partial x}{H} + \bar{v} \frac{\partial H/\partial y}{H} \right) \left(D_m^C - \delta\phi_m^C (\xi = 0) \right),$$

$$E_{mn} \frac{\partial R_n}{\partial t} - \frac{\rho_* N_{max}^2}{g} F_{mn}\,W_n = 0,$$

$$G_{mn} \frac{\partial W_n}{\partial t} + \frac{g}{\rho^*} E_{mn}\,R_n + \frac{1}{\rho^* H} H_{mn}\,P_n = 0, \qquad \begin{array}{l}\text{Summation over n,}\\ (m, n = 1, 2, 3, \ldots, N).\end{array}$$

(2.68)

The last two of these equations can also be replaced by the second order (in time) equation

$$G_{mn} \frac{\partial^2 W_n}{\partial t^2} + N_{max}^2\,F_{mn}\,W_n + \frac{H_{mn}}{\rho^* H} \frac{\partial P_n}{\partial t} = 0, \qquad \begin{array}{l}\text{Summation over n,}\\ (m, n = 1, 2, 3, \ldots, N).\end{array} \quad (2.69)$$

The various coefficient matrices are collected in *Table 2.3*. They are all expressible in terms of the inner product

$$< f(\xi),\ g(\xi) > = \int_0^1 f(\xi)\,g(\xi)\ d\xi.$$

Furthermore, $\hat{N}^2$ in *Table 2.3* is the normalized Brunt-Väisälä frequency

$$\hat{N}^2 = \frac{N^2(\xi; H)}{N_{max}^2} = -\frac{d\rho_0/dz\,(\xi, H)}{|d\rho_0/dz|_{max}}, \qquad \xi = \left(\frac{z}{H} + 1 \right).$$

The first two of equations (2.68) derive from the horizontal momentum equations (2.66)$_{1,2}$, the third corresponds to the continuity equation (2.66)$_3$ and the fourth and fifth are obtained from the adiabaticity statement $(2.66)_4$ and the vertical momentum balance $(2.66)_5$. It is also evident that (2.68) constitute 5N equations for the 5N unknowns $\{U_n, V_n, W_n, P_n, R_n\}$, provided the barotropic quantities ζ, $\bar{u}$, $\bar{v}$ are known. If they are not, equations (2.68) must be complemented by equations (2.61) which, with the use of (2.67), take on the form

$$\frac{\partial(H\bar{u})}{\partial x} + \frac{\partial(H\bar{v})}{\partial y} = 0,$$

$$\frac{\partial \bar{u}}{\partial t} - f\bar{v} + g\frac{\partial \zeta}{\partial x} = -\frac{1}{\rho*}\left(M_m \frac{\partial P_m}{\partial x} + N_m \frac{\partial H/\partial x}{H} P_m + \frac{\tau_1}{H}\right),$$

$$\frac{\partial \bar{v}}{\partial t} + f\bar{u} + g\frac{\partial \zeta}{\partial y} = -\frac{1}{\rho*}\left(M_m \frac{\partial P_m}{\partial y} + N_m \frac{\partial H/\partial y}{H} P_m + \frac{\tau_2}{H}\right).$$

(2.70)

The vectors M_m, N_m $(m=1,2,3,\ldots,N)$ are also defined in *Table 2.3*.

elmt	definition	buoyancy mode set	Jacobi polynomial set
A_{mn}^L	$<\delta\phi_m^L,\ \phi_n>,\ \ L\in\{M,C\}$	$\lambda_n\,\delta_{mn}$	δ_{mn}
B_{mn}	$<\delta\phi_m^M,\ (\xi-1)\dfrac{d\psi_n}{d\xi}> -$ $<1,(\xi-1)\dfrac{d\psi_n}{d\xi}><1,\delta\phi_m^M>$	$<\dfrac{d\Xi_m}{d\xi},\ (\xi-1)\dfrac{d^2\Xi_m}{d\xi^2}>$	$0\ (n\le m),\quad b_{n-1,m}\ (n>m)$
C_{mn}	$<\delta\phi_m^M,\ \psi_n> - <1,\psi_n><1,\delta\phi_m^M>$	$\lambda_n\,\delta_{mn}$	$\delta_{m,n-1}$
D_m^L	$<1,\delta\phi_m^L>$	0	0
E_{mn}	$<\delta\phi_m^V,\ \chi_n>$	δ_{mn}	δ_{mn}
F_{mn}	$<\delta\phi_m^V,\ \hat{N}^2\,\Xi_n>$	δ_{mn}	$<G_{m-1},\hat{N}^2 G_{n-1}>\ (=\hat{N}_0^2\,\delta_{mn},$ cont. strat.$)$
G_{mn}	$<\delta\phi_m^V,\ \Xi_n>$	$<\Xi_m,\Xi_n>$	δ_{mn}
H_{mn}	$<\delta\phi_m^V,\ \dfrac{d\psi_n}{d\xi}>$	$\lambda_n\,\delta_{mn}$	$0\ (n\le m),\quad h_{n-2,m-1}\ (n>m)$
K_{mn}	$<(\xi-1)\dfrac{d\phi_n}{d\xi},\delta\phi_m^C> - \delta\phi_m^C(0)\,\phi_n(0)$	$B_{mn} - \dfrac{d\Xi_m}{d\xi}\dfrac{d\Xi_n}{d\xi}\Big|_0$	$0\ (n<m),\ b_{nm}\ (n\ge m)\} \dfrac{-(-1)^{n+m}}{\sqrt{(2n+1)(2m+1)}}$
L_{mn}	$<\dfrac{d\,\delta\phi_m^C}{d\xi},\Xi_n>$	$\lambda_n\,\delta_{mn}$	$h_{m-1,n-1}\ (n\le m),\quad 0\ (n>m)$
M_m	$<1,\psi_m>$	0	δ_{1m}
N_m	$<1,\psi_m> - \psi_m(0)$	$-\dfrac{d\Xi_m}{d\xi}\Big|_{\xi=0}$	$\delta_{1m} + (-1)^m\sqrt{2m-1}$

Table 2.3 Matrix elements for the expansion of the field variables in terms of the flat-bottom buoyancy modes or Jacobi polynomials. $<a,b>$ is the inner product $\int_0^1 ab\,d\xi$, λ_n is the eigenvalue defined in (2.71).

It is our contention that by accordingly selecting the shape functions the barotropic modes and the baroclinic modes can almost completely be separated. This orthogonalization is exactly possible in stratified basins with constant depth; selecting the shape functions from this set will nearly achieve the uncoupling in the case of a variable bottom. Towards a motivation, consider equations (2.64) and ignore the vertical acceleration terms, i.e. restrict considerations to quasi-static pressure conditions. For this case (2.64) may be reduced to the single partial differential equation for w

$$N^2 \, \nabla^2 \, w + (\frac{\partial^2}{\partial t^2} + f^2) \, \frac{\partial^2 \, w}{\partial z^2} = 0.$$

Subject to the boundary conditions

$$w = 0, \quad \text{at} \quad z = 0, -H$$

this equation permits separation of variable solutions $w(x,y,z,t) = W_n(x,y,t) \cdot Z_n(z)$, where $Z_n(z)$ satisfies the eigenvalue problem

$$Z_n''(z) + \frac{N^2(t)}{g \, H_n} \, Z_n(z) = 0, \quad -H < z < 0,$$

$$Z_n = 0, \quad z = -H, \, 0,$$

with the eigenvalue $g H_n$. Introducing the transformation $z = H \cdot (\xi - 1)$ this becomes

$$\frac{d^2 \, Z_n(z)}{d\xi^2} + \lambda_n \, \hat{N}^2(\xi; H) \, Z_n(\xi) = 0, \quad 0 < \xi < 1, \quad (2.71\text{a})$$

$$Z_n(\xi) = 0, \quad \xi = 0, \, 1, \quad (2.71\text{b})$$

$$\lambda_n = \frac{N_{max}^2 \, H^2}{g \, H_n}, \quad (2.71\text{c})$$

where λ_n is the eigenvalue. This is the classical eigenvalue problem of internal waves in a basin of constant depth. It is selfadjoint, and so λ_n is real and positive for all $n = 1, 2, \ldots$; furthermore, the eigenfunctions form a complete set and can be normalized to satisfy the orthogonality relations

$$< \hat{N}^2(\xi; H) \, Z_m(\xi), \, Z_n(\xi) > = \delta_{mn}. \quad (2.72)$$

We conjecture that by selecting shape functions and weighting functions from this set or from derivatives of it we will achieve a weak coupling of the essentially barotropic-TW motion with the internal wave motion. The arguments are:

(i) If we choose $\Xi_n = Z_n$ the vertical velocity profiles are those of the internal wave motion of a fluid with constant depth (Figure 2.4). Actual boundary conditions at the bottom are not satisfied by these functions. This will result in a coupling of the different internal modes.

(ii) If we further select $\phi_m = d\,\Xi_m/d\xi$ we will exactly match the vertical distribution of the horizontal velocity profiles for internal waves in a basin with constant depth. To be consistent with (2.65b) the function set $\{\phi_m\}$ must be orthogonal to the constant function. One can easily verify that

$$< 1, \phi_m > \; = \; < 1, \frac{d\Xi_m}{d\xi} > \; = \; \int_0^1 \frac{dZ_m}{d\xi} \; d\xi \; = \; Z_m(1) - Z_m(0) \; = \; 0,$$

by virtue of the boundary condition in (2.71b).

(iii) The momentum equations $(2.64)_{1,2,5}$, the continuity equation $(2.64)_3$ and the adiabaticity equation $(2.64)_4$ now suggest that we should choose

$$\{\psi_m\} \; = \; \{\phi_m\} \; = \; \{d\Xi_m/d\xi\} \quad \text{and} \quad \{\chi_m\} \; = \; \{\hat{N}^2 \, \Xi_m\}.$$

(iv) We weight the horizontal momentum equations with the same weighting function as the shape functions of horizontal velocity. Similarly, the shape functions and the weighting functions in the continuity equation should be chosen from the same function set. This yields

$$\{\delta\phi_m^M\} \; = \; \{\delta\phi_m^C\} \; = \; \{\frac{d\Xi_m}{d\xi}\}.$$

Finally, the adiabatic equation and the vertical momentum equation then suggest that

$$\{\delta\phi_m^V\} \; = \; \{\Xi_m\}.$$

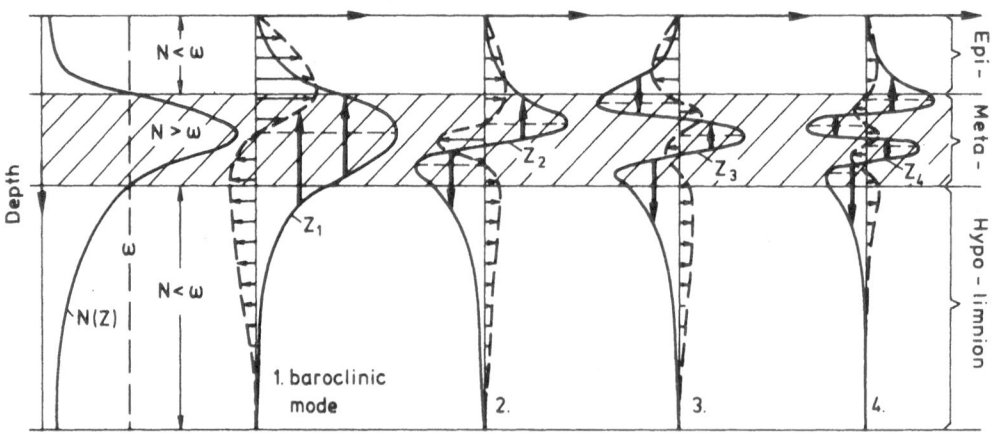

Figure 2.4

Typical vertical distribution of the Brunt-Väisälä frequency N (left) and the four lowest baroclinic modes (qualitative). Solid curves show the distribution of the vertical velocity component, dashed curves indicate the distribution of the longitudinal velocity component (when f = 0), Most energy is usually concentrated in the first baroclinic mode, the exact distribution must, however, be determined by continuous profiles of horizontal velocities. [From Hutter, 1984b]

With these choices the matrix elements can readily be calculated; they are listed in column 3 of *Table 2.3*. This table also gives the elements for an alternative selection of basis functions. Leading idea in postulating (2.72) was to incorporate into the function set as much as possible of the particular physics under consideration. Consequently, the complete function set was that of the eigenfunctions for buoyancy waves in a stratified basin of constant depth. From a computational point of view this approach implies that the eigenvalue problem (2.71) must be solved in advance in order to obtain the required basis functions Ξ_n. Alternatively, we could expand the functions $\tilde{u}$, $\tilde{v}$, w, p and ρ in terms of special *orthogonal polynomials*. Indeed, the scalar product $< \theta, \psi > = \int_0^1 \theta\psi d\xi$ suggests the use of Jacobi polynomials $G_n(1, 1, \xi)$ (see Abramowitz & Stegun, 1972), which are orthonormal in the interval $[0,1]$ with respect to the weighting function 1 in the scalar product. They are defined by

$$G_n(1, 1, \xi) = \frac{\sqrt{2n+1}}{n!} \sum_{k=0}^{n} (-1)^{n-k} \binom{n}{k} \frac{(n+k)!}{k!} \xi^k ,$$

$$G_0 = 1,$$

$$G_1 = \sqrt{12} \left(\xi - \frac{1}{2}\right),$$

$$G_2 = \sqrt{180} \left(\xi^2 - \xi + \frac{1}{6}\right), \ldots$$

and satisfy the orthogonality relations

$$\int_0^1 G_n \cdot G_m \, d\xi = < G_n, G_m > = \delta_{mn}.$$

Note that the constant function G_0 is the first basis function and all G_n for $n > 0$ are orthogonal to it, i.e. they span a vanishing vertical area. This is well in accord with the condition (2.65b). The derivative of G_{n+1} is a polynomial of degree n and can be expressed as a linear combination of G_k, $k = 1, 2, \ldots, n$. For later use we define

$$(\xi - 1) \frac{dG_n}{d\xi} \equiv \sum_{k=0}^{n} b_{nk} G_k ,$$

$$\frac{dG_{n+1}}{d\xi} \equiv \sum_{k=0}^{n} h_{nk} G_k .$$

The advantage of this polynomial set is its easy accessibility that frees us from solving a problem oriented eigenvalue problem. Its likely disadvantage is a slower convergence in comparison to the "physical set" of internal eigenfunctions.

We now select

$$\{\phi_m\} = \{G_m\},$$

$$\{\Xi_m\} = \{\psi_m\} = \{\chi_m\} = \{G_{m-1}\},$$

$$m = 1,\ldots,N$$

and the weighting function

$$\{\delta\phi_m^M\} = \{\delta\phi_m^C\} = \{G_m\},$$

$$\{\delta\phi_m^V\} = \{G_{m-1}\},$$

$$m = 1,\ldots,N$$

The corresponding matrix elements are listed in column 4 of *Table 2.3*. We now see the distinct properties of the two alternative approaches: When non-hydrostatic terms are ignored an expansion in terms of buoyancy modes uncouples the individual baroclinic modes in equation (2.69). The barotropic-baroclinic coupling arises in the horizontal momentum equations $(2.68)_{1,2,3}$ and $(2.70)_{2,3}$ only in conjunction with topographic gradients. For a flat bottom all barotropic and baroclinic modes are uncoupled. On the other hand, the Jacobi set, even though it is more easily accessible does not follow the physics so closely. Notice also that the buoyancy equations (involving E_{mn} and F_{mn}) are strongly coupled in this case whereas the coupling due to topography gradients in the horizontal momentum equations (involving B_{mn}) is weaker.

Summarizing the governing equations we obtain *for an expansion in buoyancy modes* only:

$$\frac{\partial(H\bar{u})}{\partial x} + \frac{\partial(H\bar{v})}{\partial y} = 0,$$

$$\frac{\partial\bar{u}}{\partial t} - f\bar{v} + g\frac{\partial\zeta}{\partial x} = -\frac{1}{\rho_*}\frac{d\Xi_m}{d\xi}(0)\frac{\partial H/\partial x}{H}P_m + \frac{\tau_1}{\rho_* H},$$

$$\frac{\partial\bar{v}}{\partial t} + f\bar{u} + g\frac{\partial\zeta}{\partial y} = -\frac{1}{\rho_*}\frac{d\Xi_m}{d\xi}(0)\frac{\partial H/\partial y}{H}P_m + \frac{\tau_2}{\rho_* H},$$

$$\frac{\partial U_m}{\partial t} - fV_m + \frac{1}{\rho_*}\frac{\partial P_m}{\partial x} - \frac{1}{\rho_*}\lambda_m^{-1}B_{m\ell}\frac{\partial H/\partial x}{H}P_\ell = 0,$$

$$\frac{\partial V_m}{\partial t} + fU_m + \frac{1}{\rho_*}\frac{\partial P_m}{\partial y} - \frac{1}{\rho_*}\lambda_m^{-1}B_{m\ell}\frac{\partial H/\partial y}{H}P_\ell = 0,$$

$$\frac{\partial U_m}{\partial x} + \frac{\partial V_m}{\partial y} - \frac{W_m}{H} - \lambda_m^{-1}K_{m\ell}\left[\frac{\partial H/\partial x}{H}U_\ell + \frac{\partial H/\partial y}{H}V_\ell\right]$$

$$= -\lambda_m^{-1}\frac{d\Xi_m}{d\xi}(0)\left[\frac{\partial H/\partial x}{H}\bar{u} + \frac{\partial H/\partial y}{H}\bar{v}\right],$$

$$G_{mn}\frac{\partial^2 W_n}{\partial t^2} + N_{max}^2 W_m + \frac{1}{\rho_* H}\lambda_m\frac{\partial P_m}{\partial t} = 0,$$

(2.73)

$$(m,n,\ell = 1,2,3\ldots,N).$$

Boundary conditions which must be satisfied along the shore are

$$\bar{\underset{\sim}{u}} \cdot \underset{\sim}{n} = 0,$$

$$U_m \cdot \underset{\sim}{n} = 0, \quad m = 1, 2, \ldots, N, \qquad \text{along the shore} \qquad (2.74)$$

where $\underset{\sim}{n}$ is the unit normal vector and it is assumed that the depth H does not vanish along the shore. The wavy underlined terms in (2.73) describe the barotropic-baroclinic coupling. All these terms involve the gradient of H.

b) *Scale analysis*

To estimate the significance of the barotropic-baroclinic coupling, let us non-dimensionalize equations (2.73). To this end we introduce the following scales and dimensionless variables:

$$(x, y, H) = ([L] x*, [L] y*, [H] H*),$$

$$\zeta = [\zeta] \zeta*, \qquad t = [f^{-1}] t*,$$

$$(\bar{u}, \bar{v}, U_m, V_m) = [U](\bar{u}*, \bar{v}*, U_m^*, V_m^*),$$

$$W_m = [W] W_m^*, \qquad (2.75)$$

$$P_m = [P] P_m^*$$

$$\underset{\sim}{\tau}^{wind} = [\tau^{wind}] \underset{\sim}{\tau}*.$$

Bracketed quantities are orders of magnitude of the variables in question and variables having an asterisk are dimensionless. With (2.75), (2.73) become (asterisks are consistently omitted):

$$\frac{\partial}{\partial x}(H\bar{u}) + \frac{\partial}{\partial x}(H\bar{v}) = 0,$$

$$\frac{\partial \bar{u}}{\partial t} - \bar{v} + \mathbb{A} \frac{\partial \zeta}{\partial x} = - \mathbb{B} \frac{d\Xi_m(0)}{d\xi} \frac{\partial H/\partial x}{H} P_m + \mathbb{C} \frac{\tau_1}{H},$$

$$\frac{\partial \bar{v}}{\partial t} + \bar{u} + \mathbb{A} \frac{\partial \zeta}{\partial y} = - \mathbb{B} \frac{d\Xi_m(0)}{d\xi} \frac{\partial H/\partial y}{H} P_m + \mathbb{C} \frac{\tau_2}{H},$$

$$\frac{\partial U_m}{\partial t} - V_m + \mathbb{B} \frac{\partial P_m}{\partial x} - \mathbb{B} \lambda_m^{-1} B_{m\ell} \frac{\partial H/\partial x}{H} P_\ell = 0,$$

$$\frac{\partial V_m}{\partial t} + U_m + \mathbb{B} \frac{\partial P_m}{\partial y} - \mathbb{B} \lambda_m^{-1} B_{m\ell} \frac{\partial H/\partial y}{H} P_\ell = 0, \qquad (2.76)$$

$$\frac{\partial U_m}{\partial x} + \frac{\partial V_m}{\partial y} - \mathbb{D} \frac{W_m}{H} - \lambda_m^{-1} K_{m\ell} \left[\frac{\partial H/\partial x}{H} U_\ell + \frac{\partial H/\partial y}{H} V_\ell \right]$$

$$= - \lambda_m^{-1} \frac{d\Xi_m(0)}{d\xi} \left[\frac{\partial H/\partial x}{H} \bar{u} + \frac{\partial H/\partial y}{H} \bar{v} \right],$$

$$\mathbb{E} \, G_{mn} \frac{\partial^2 W_n}{\partial t^2} + W_m + \mathbb{F} \, \lambda_m \frac{1}{H} \frac{\partial P_m}{\partial t} = 0,$$

where

$$A = \frac{g[\zeta]}{f[L][U]}, \qquad \mathbb{C} = \frac{[\tau]}{\rho* f[H][U]}, \qquad \mathbb{E} = \frac{f^2}{N_{max}^2},$$

$$B = \frac{[P]}{\rho* f[L][U]}, \qquad \mathbb{D} = \frac{[L][W]}{[H][U]}, \qquad \mathbb{F} = \frac{f[P]}{\rho* [H][W] N_{max}^2}. \tag{2.77}$$

Choosing the scales according to[*)]

$$[L] = 10^4 \text{ m}, \qquad [H] = 10^2 \text{ m}, \qquad [\zeta] = 10^{-1} \text{ m},$$

$$[f^{-1}] = 10^4 \text{ s}, \qquad [U] = 1 \text{ m s}^{-1}, \qquad [W] = 10^{-2} \text{ m s}^{-1},$$

$$[P/\rho*] = 10^{-1} \text{ m}^2 \text{ s}^{-2}, \quad [\tau/\rho*] = 10^{-2} \text{ m}^2 \text{ s}^{-2}, \quad N_{max}^2 = 10^{-3} \text{ s}^{-2},$$

the orders of magnitude of (2.77) are

$$A = O(1), \qquad \mathbb{C} = O(1), \qquad \mathbb{E} = O(10^{-5}),$$
$$B = O(10^{-1}), \qquad \mathbb{D} = O(1), \qquad \mathbb{F} = O(1).$$

Important in the following argument are only the values for A, B and $\mathbb{C}$. Thus the barotropic-baroclinic coupling term (wavy underlined) in the momentum equations of the barotropic motion is small in comparison to the remaining terms of this equation, but this cannot be said about the baroclinic-barotropic coupling term (wavy underlined) in the baroclinic continuity equation, because this term does not contain a factor B while at least some of the remaining terms in the equation are order unity[**)]. This argument demonstrates that the barotropic-baroclinic coupling is weak in the sense that to lowest order the barotropic motion is unaffected by the baroclinic processes. On the other hand a baroclinic trace of the barotropic motion can be discerned, because to the same order of accuracy the barotropic flow serves as an input to the baroclinic response.

This is then the approximate solution procedure: We solve in a first step the TW-equation

$$\nabla \cdot (\frac{\nabla \psi_t}{H}) + J(\psi, \frac{f}{H}) = 0, \qquad \text{in } \mathcal{D}$$

$$\psi = 0, \qquad \text{on } \partial\mathcal{D}$$

evaluate $\bar{u}$ and $\bar{v}$ according to (2.62) and substitute them into $(2.76)_6$.

---

*) An estimate for $[P/\rho*]$ is obtained as follows: Under hydrostatic conditions the last of equations suggests that $[P/\rho*] \sim (\Delta\rho/\rho)g [D]$ where $\Delta\rho/\rho$ is the density anomaly and $[D]$ a typical metalimnion thickness: Thus with $\Delta\rho/\rho \sim 10^{-3}$ and $[D] \lesssim 10$ m this yields $[P/\rho*] \lesssim 10^{-1}$ m² s⁻², implying $B \lesssim 10^{-1}$.

**) This argument can even be made more forceful by recognizing that according to (2.71c) an estimate for H_1 is 10 m ($n = 1$) so that $\lambda_1 = 10^{-3} \times 10^4/10^2 = 10^{-1}$. Consequently one baroclinic-barotropic coupling term is about a factor of 100 larger than the corresponding barotropic-baroclinic term.

The internal wave problem (equations $(2.76)_{4,5,6,7}$) is then solved in a second step. Structurally, this is analogous to the two-layer case studied before.

2.5 TW-equation in orthogonal coordinate systems

In chapter 3 the TW-equation (2.22) will be discussed in various coordinate systems. To keep technical details short, these are summarized in this section.

a) *Preparation*

We restrict our considerations to *orthogonal* coordinate systems $\underset{\sim}{x} = (x_1, x_2, x_3)$ which have the property that their metric tensor $\underset{\approx}{g}$ has diagonal form, e.g. in $\mathbb{R}^3$

$$\underset{\approx}{g} = \begin{bmatrix} J_1 & 0 & 0 \\ 0 & J_2 & 0 \\ 0 & 0 & J_3 \end{bmatrix} . \tag{2.78}$$

The arc element $d\underset{\sim}{\ell}$ can be expressed using

$$d\underset{\sim}{\ell} = \underset{\approx}{g} \, d\underset{\sim}{x} , \tag{2.79}$$

where $d\underset{\sim}{x} = (dx_1, dx_2, dx_3)$ is the increment vector and insertion of (2.78) into (2.79) yields

$$d\underset{\sim}{\ell} = (J_1 \, dx_1, \, J_2 \, dx_2, \, J_3 \, dx_3) . \tag{2.80}$$

Table 2.4 collects the components of the metric tensor for frequently used orthogonal coordinate systems.

Coordinates (x_1, x_2, x_3)		J_1	J_2	J_3	
Cartesian	(x, y, z)	1	1	1	
cylindric	(r, ϕ, z)	1	r	1	
elliptic	(ξ, η, z)	J	J	1	, where $J = a(\sinh^2 \xi + \sin^2 \eta)^{1/2}$
natural	(s, n, z)	J	1	1	, where $J = 1 - K(s) \cdot n$

Table 2.4   Coordinate systems often used in lake hydrodynamics.

We recall the TW-equation (2.22) and the barotropic velocity field in the coordinate-invariant formulation

$$\frac{\partial}{\partial t}(\nabla \cdot (\frac{\nabla \psi}{H})) + \hat{\underset{\sim}{z}} \cdot (\nabla \psi \times \nabla \frac{f}{H}) = 0,$$

$$\underset{\sim}{u} = \frac{1}{H}(\hat{\underset{\sim}{z}} \times \nabla \psi) . \tag{2.22}$$

To obtain these equations in the different coordinate systems the vector differential operators need be written in curvilinear coordinates. In the orthogonal coordinate system whose arc element has the form (2.80) the gradient-, divergence- and curl-operator are given by

$$\text{grad } u = \left(\frac{1}{J_1}\frac{\partial u}{\partial x_1}\,;\, \frac{1}{J_2}\frac{\partial u}{\partial x_2}\,;\, \frac{1}{J_3}\frac{\partial u}{\partial x_3}\right),$$

$$\text{div } \underset{\sim}{v} = \frac{1}{J_1 J_2 J_3}\left(\frac{\partial}{\partial x_1}(J_2 J_3 v_1) + \frac{\partial}{\partial x_2}(J_1 J_3 v_2) + \frac{\partial}{\partial x_3}(J_1 J_2 v_3)\right),$$

$$\text{curl } \underset{\sim}{v} = \frac{1}{J_1 J_2 J_3}\left[J_1\frac{\partial}{\partial x_2}(J_3 v_3) - J_1\frac{\partial}{\partial x_3}(J_2 v_2)\,;\right.$$

$$J_2\frac{\partial}{\partial x_3}(J_1 v_1) - J_2\frac{\partial}{\partial x_1}(J_3 v_3)\,;$$

$$\left. J_3\frac{\partial}{\partial x_1}(J_2 v_2) - J_3\frac{\partial}{\partial x_2}(J_1 v_1)\right],$$

(2.81)

where $u = u(\underset{\sim}{x})$ is a scalar and $\underset{\sim}{v} = \underset{\sim}{v}(\underset{\sim}{x})$ is a vector in this coordinate system. A derivation is found in Pearson (1974).

b) Cylindrical coordinates

These coordinates are often used in problems which exhibit some rotational symmetry. The coordinates are (r, ϕ, z) which are related to the Cartesian system by the formulae

$$\begin{aligned} x &= r\cos\phi, \\ y &= r\sin\phi, \qquad r \geq 0;\ 0 \leq \phi \leq 2\pi \\ z &= z. \end{aligned}$$

The arc element is given by

$$d\underset{\sim}{\ell} = (dr,\ r\,d\phi,\ dz),$$

as anticipated in *Table 2.4*. Applying (2.81) to (2.22) yields

$$\left(\frac{r}{H}\,\psi_{tr}\right)_r + \left(\frac{1}{rH}\,\psi_{t\phi}\right)_\phi + \psi_r\left(\frac{f}{H}\right)_\phi - \psi_\phi\left(\frac{f}{H}\right)_r = 0.$$

(2.82)

c) Elliptical coordinates

The coordinates of the elliptic cylinder system are (ξ, η, z), and for fixed z the lines $\xi = $ const are confocal ellipses whereas the lines $\eta = $ const are hyperbolas, see *Figure 2.5*. The parameter a denotes the position of the foci, and the Cartesian coordinates are calculated from (ξ, η, z) by

$$\begin{aligned} x &= a\cosh\xi\cos\eta, \\ y &= a\sinh\xi\sin\eta, \qquad \xi \geq 0,\ 0 \leq \eta \leq 2\pi \\ z &= z. \end{aligned}$$

The shore line of the elliptic basin is given by

$$\frac{x^2}{(a \cosh \xi_S)^2} + \frac{y^2}{(a \sinh \xi_S)^2} = 1,$$

which is an ellipse with the semi-axes A and B and an aspect ratio (width to length)

$$r = \frac{B}{A} = \frac{a \sinh \xi_S}{a \cosh \xi_S} = \tanh \xi_S. \tag{2.83}$$

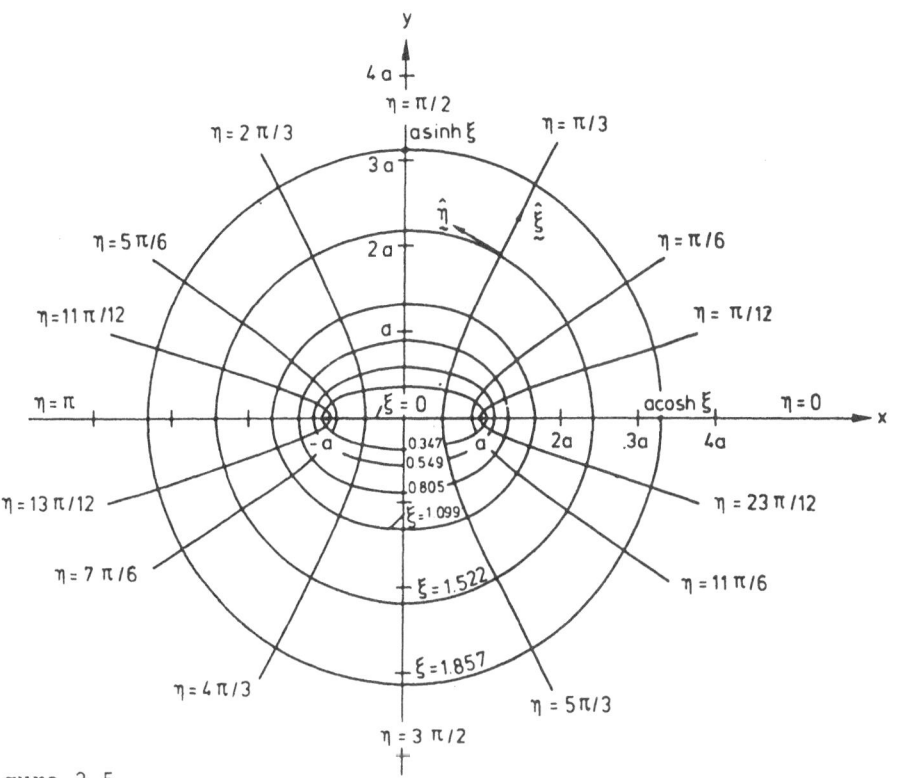

Figure 2.5
Elliptic cylinder coordinates (ξ, η). The quantities $\hat{\xi}$ and $\hat{\eta}$ are unit vectors in the directions of increasing ξ and η. We refer to ξ and η as radial and angular coordinates respectively. [From Mysak, 1983]

The first two diagonal elements of the metric tensor are equal and read

$$J \equiv J_1 = J_2 = a(\sinh^2 \xi + \sin^2 \eta)^{1/2},$$

$$J_3 = 1.$$

Thus the TW-equation in elliptic coordinates takes the form

$$\left(\frac{1}{H} \psi_{t\xi}\right)_\xi + \left(\frac{1}{H} \psi_{t\eta}\right)_\eta + \psi_\xi \left(\frac{f}{H}\right)_\eta - \psi_\eta \left(\frac{f}{H}\right)_\xi = 0. \tag{2.84}$$

Due to the equality $J \equiv J_1 = J_2$ the equation for the streamfunction is independent of J so that, formally, the same equation as in a Cartesian system is obtained. However, the metric factor enters the formula for the velocity field

$$\underset{\sim}{u} = (- \frac{1}{JH} \psi_n, \frac{1}{JH} \psi_\xi). \qquad (2.85)$$

d) *Natural coordinates*

For the developments in subsequent chapters we need the TW-equation also in a natural coordinate system. With this, it is particularly convenient to describe elongated and curved lake basins. We choose an orthogonal network which spans the elongated domain. The basis for it is an axis, which follows more or less the *thalweg*[*)] of the lake. The arc length s along this axis forms the first coordinate of the system.

In view of the restriction to elongated narrow lakes it is possible to choose a straight n-axis; so the system is curved only in the s-direction, see *Figure 2.6*. In order to define the lake domain uniquely in terms of these coordinates the radius of curvature R(s) must exceed half the

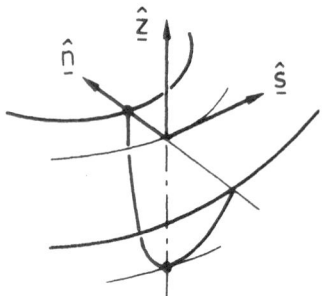

Figure 2.6

The natural coordinate system (s,n,z) in the lake basin with unit vectors $\hat{s}$, $\hat{n}$ and $\hat{z}$. $\hat{n}$ points to the positive center of the curvature along s.

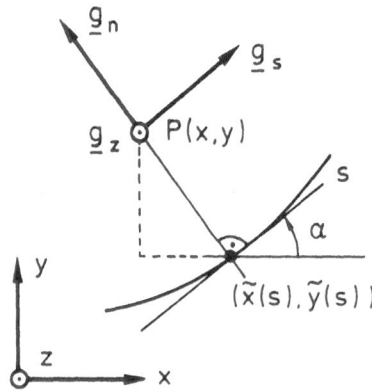

Figure 2.7 Basis vectors in the natural coordinate system.

width of the lake B(s), R(s) > B(s)/2. Let the lake axis be given by a parameter representation ($\tilde{x}(s)$, $\tilde{y}(s)$) within a Cartesian system as shown in *Figure 2.7*. The coordinates of an arbitrary point P are then given by

$$x = \tilde{x}(s) - n \sin\alpha(s), \qquad y = \tilde{y}(s) + n \cos\alpha(s),$$

provided the n-axis is chosen to be straight. The set of basis vectors $\underset{\sim}{g}_s$, $\underset{\sim}{g}_n$ and $\underset{\sim}{g}_z$ at point P can be expressed in the form

$$\underset{\sim}{g}_s = (\frac{dx}{ds}, \frac{dy}{ds}, 0), \qquad \underset{\sim}{g}_n = (\frac{dx}{dn}, \frac{dy}{dn}, 0), \qquad \underset{\sim}{g}_z = (0, 0, 1),$$

[*)] *The thalweg of an elongated lake is the line which follows the deepest points of the basin cross sections.*

which is easily simplified to the form

$$\underset{\sim}{g}_s = (\tilde{x}' - n K \cos\alpha, \tilde{y}' - n K \sin\alpha, 0).$$

$$\underset{\sim}{g}_n = (\quad -\sin\alpha\quad,\quad \cos\alpha\quad, 0),$$

$$\underset{\sim}{g}_z = (\quad 0\quad,\quad 0\quad, 1),$$

using as definition for the curvature $K = d\alpha/ds$.

Figure 2.8

Arc element in a natural
coordinate system.

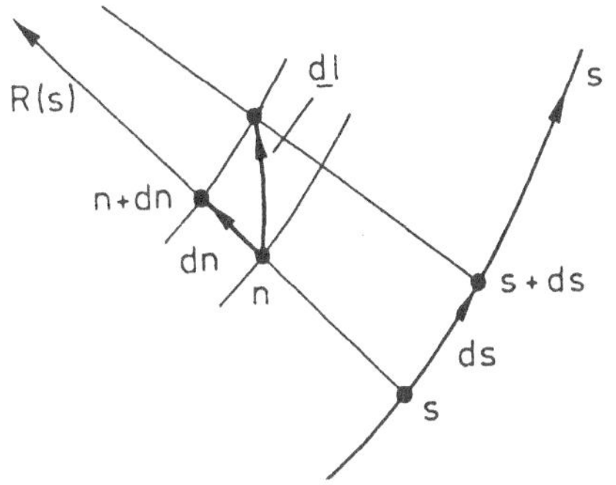

With the aid of *Figure 2.8* it follows that the arc element $d\underset{\sim}{\ell}$ takes the form

$$d\underset{\sim}{\ell} = (J\ ds, dn, dz), \qquad J = 1 - Kn.$$

We finally obtain the TW-equation and the velocity field in natural co-ordinates:

$$\left(\frac{1}{JH}\ \psi_{ts}\right)_s + \left(\frac{J}{H}\ \psi_{tn}\right)_n + \psi_s \left(\frac{f}{H}\right)_n - \psi_n \left(\frac{f}{H}\right)_s = 0 ,$$

$$\underset{\sim}{u} = \left(-\frac{1}{H}\ \psi_n , \ \frac{1}{JH}\ \psi_s\right).$$

(2.86)

e) *Cartesian-coordinate correspondence principle*

It was mentioned that the TW-equation in elliptical coordinates is for-mally the same as that in Cartesian coordinates. Because also the bound-ary conditions of no flux are the same a correspondence principle can be applied to construct solutions in the elliptical coordinates from those that are already known in Cartesian coordinates. Of course, for the correspondence principle to apply the bathymetric functions in the two systems must also correspond.

For instance, TW-solutions in hyperbolic channels can easily be deduced from corresponding solutions in straight channels (see chapter 5) and TW's in basins which are bounded by confocal ellipses and hyperbolas can be deduced from corresponding TW-solutions in rectangles (chapter 6).

3. Some known solutions of the TW-equation in various domains

Whereas our intentions in chapter 1 have been to suggest that observations of long-periodic oscillations in lakes and ocean basins are likely to be interpretable in terms of vorticity induced motions, we have presented in chapter 2 the basic equations of such waves and analysed the general properties of the associated equations.

In this and ensuing chapters we will construct explicit solutions. Our principal aim is to extract through this analysis the physical properties of TW's and to see in what respect the interpretations surmised in the first chapter can be substantiated. Many solutions are known from the literature, but we also include new results.

We shall discuss (i) circular basins with a topography being a function of the radial distance only (parabola, power-law), (ii) elliptic basins with a parabolic bottom and an exponential shelf profile, (iii) infinite channels and shelves and, (iv) TW's around an elliptical island. All these domains are characterized by the fact that the isobaths follow one coordinate line of the coordinate system, so that ordinary differential equations emerge. As a result the mathematical tool is solving two-point boundary-value problems.

3.1 Circular basin with parabolic bottom

Following Lamb (1932, § 212) we start our analysis of TW's in circular basins with equation (2.29), or

$$\nabla \cdot (h \nabla \zeta_t) + J(h, \zeta) - (\frac{L}{R})^2 L \zeta_t = 0. \tag{3.1}$$

Here all quantities are dimensionless except L and R, a typical length and the Rossby radius, respectively[*]. In polar coordinates (3.1) may be written as

$$(h \zeta_{rt})_r + \frac{1}{r} (\frac{h}{r} \zeta_{\theta t})_\theta + \frac{h}{r} \zeta_{rt} +$$

$$+ \frac{h_r}{r} \zeta_\theta - \frac{h_\theta}{r} \zeta_r - (\frac{L}{R})^2 L \zeta_t = 0, \qquad\qquad 0 < r < 1. \tag{3.2}$$

The boundary conditions (no mass flux at the outer boundary, finiteness of ζ at the origin) are

---

[*] *For a subtlety in defining the Rossby radius see footnote on p. 34.*

$$\zeta = \text{finite}, \qquad \text{at} \quad r = 0,$$
$$\zeta_r = 0, \qquad \text{at} \quad r = 1. \tag{3.3}$$

Consider a radial topography,

$$h = h(r) \tag{3.4}$$

and assume an azimuthal wave solution of the form

$$\zeta = Z(r) \exp[i(m\theta - \sigma t)], \tag{3.5}$$

travelling counterclockwise around the basin; $\sigma = \omega/f$ is the dimensionless frequency and m the azimuthal wavenumber. With (3.4) and (3.5), the boundary value problem (3.2)-(3.3) assumes the form

$$(h Z')' + \frac{h}{r} Z' - \left[\frac{m^2}{r^2} h + \frac{m}{\sigma} \frac{h'}{r} + \frac{1-\sigma^2}{(R/L)^2} \right] Z = 0, \qquad 0 < r < 1, \tag{3.6}$$

$$Z = \text{finite}, \quad r = 0; \quad Z' = 0, \quad r = 1.$$

Primes denote differentiations with respect to r. From these equations the solutions presented by Lamb (1932), Wenzel (1978) and Saylor, Huang & Reid (1980) can be obtained as special cases.

For the parabolic bottom profile,

$$h = 1-r^2,$$

Z(r) can be expressed in terms of a hypergeometric polynomial (Lamb, 1932, § 212; Miles & Ball, 1963; Abramowitz & Stegun, 1972)

$$Z(r) = A_{mj} r^m F(m+j; 1-j; m+1; r^2), \qquad \begin{array}{l} m = 0,1,2,\ldots, \\ j = 1,2,3,\ldots, \end{array} \tag{3.7}$$

in which A_{mj} is a free amplitude and σ satisfies the frequency relation

$$\frac{\sigma^2-1}{(R/L)^2} + \frac{2m}{\sigma} = 2[2j(m+j-1)-m], \qquad \begin{array}{l} m = 0,1,2,\ldots, \\ j = 1,2,3,\ldots. \end{array} \tag{3.8}$$

The frequency occurs in third order which corresponds to three wave types, two first class and one second class wave. Here, we concentrate on second class waves and will therefore exclude the case m = 0. (3.8) is then equivalent to

$$\frac{1}{\sigma} = \{ \frac{2j(m+j-1)}{m} - 1 \} - \frac{\sigma^2-1}{2m(R/L)^2}.$$

The last term on the rhs represents the influence of the size effect via the external Rossby radius R and a length scale L. An order of magnitude for R is 500 km and an upper bound for L may be 200 km (Great Lakes), so $2(R/L)^2 \gtrsim 12$. The minimum value of the term in curly brackets is 1, which suggests that the two first class modes entering via the size de-

pendent term may be suppressed. Approximately, we may write, after neg-
lection of the inertial motion ($\sigma = 1$ for $j = 1$) and transformation $n = j-2$,
$n = 0, 1, \ldots$

$$\frac{1}{\sigma} = \frac{2(n+2)(m+n+1)}{m} - 1, \quad \begin{matrix} m = 1,2,\ldots, \\ n = 0,1,\ldots, \end{matrix} \tag{3.9}$$

We thus obtain the approximate frequencies and periods of *Table 3.1*. The
real parts of the surface elevation ζ and the mass transport stream
function ψ for the mode $(m, n) = (1, 0)$ are given by

$$\zeta(r,t) = Ar(1-\frac{3}{2}r^2)\cos(\theta-\sigma t),$$

$$\psi(r,t) = Ar(1-r^2)(1-\sigma-\frac{3}{2}(1-3\sigma)r^2)\cos(\theta-\sigma t).$$

The streamlines of this solution and those of the $(1,1)$- and $(2,0)$- modes
are sketched in *Figure 3.1*. The periods are larger than 50.7 hours (2.1

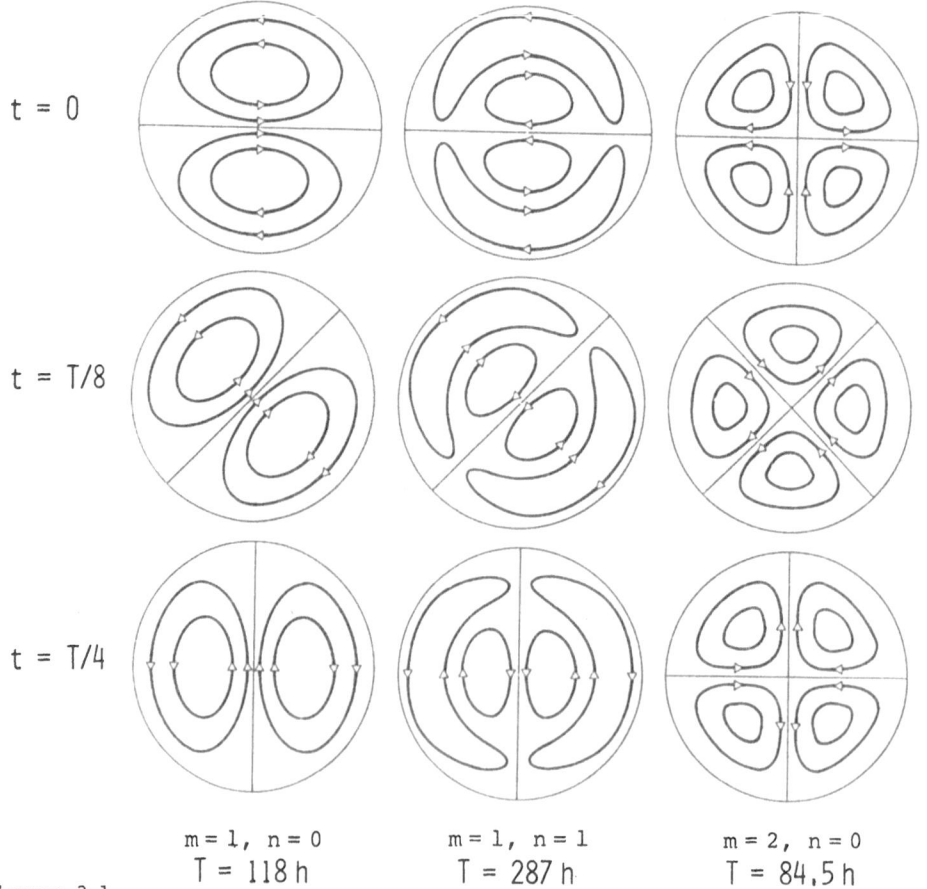

$$t = 0$$

$$t = T/8$$

$$t = T/4$$

| m = 1, n = 0 | m = 1, n = 1 | m = 2, n = 0 |
| T = 118 h | T = 287 h | T = 84,5 h |

Figure 3.1
Contour lines of the mass transport stream function of the three modes
with the simplest wave structure. The gyres rotate anticlockwise (on the
Northern hemisphere) around the basin.

days), but it is likely that only those modes with a very simple structure will be excited. The simplest fundamental mode has a period of 118 hours and consists of two basin-wide gyres, a cyclonic and an anticyclonic vortex. The entire system of gyres rotates counterclockwise (on the Northern hemisphere) around the basin. From this

m	n = 0 σ	n = 0 T[h]	n = 1 σ	n = 1 T[h]	n = 2 σ	n = 2 T[h]
1	0.143	118	0.0588	287	0.323	524
2	0.200	84.5	0.0909	186	0.0526	321
4	0.250	67.6	0.125	135	0.0769	220
∞	0.333	50.7	0.200	84.5	0.143	118

Table 3.1
Periods of TW's in a circular basin with parabolic bottom profile ignoring the size dependent term in (3.8) and computed with $f = 2\pi/16.9$ h.

fact it follows that current vectors at mid basin positions rotate in the anticlockwise direction while those at near-shore positions rotate in the clockwise direction. This is exactly the current pattern discovered by Saylor, Huang & Reid (1980) in Southern Lake Michigan, but the period of 118 hours is too large to fit the period of 100 hours inferred from measurements. They therefore studied the effect of the topography on the TW-solutions in a circular basin.

3.2 Circular basin with a power-law bottom profile

The following analysis is due to Saylor, Huang & Reid (1980) which investigated the influence of topography gradients on the periods of topographic wave motion. They used the profile

$$h(r) = (1-r^q), \qquad 0 \leq r \leq 1, \qquad q > 0. \qquad (3.10)$$

Varying the exponent q yields an entire sequence of profile geometries with strong and weak topography gradients. For q = 1 the radial depth profile is conical, for q = 2 it is parabolic, for q > 2 it becomes blunt and for q → ∞ approaches constant depth. On the other hand, for 0 < q ≤ 1 the profile has a vertex at the center and (except for q = 1) has a convex curvature similar to the exponential profile often used in shelf wave analysis. With $\psi = \psi(r) \exp(i(m\theta - \sigma t))$, use of the TW-equation in cylindrical coordinates (2.82) and a trial solution

$$\psi(r) = A r^m h^2(r), \qquad (3.11)$$

(which satisfies the boundary conditions) the depth profile must fulfil the differential equation

$$h'' + \frac{3m+2-\dfrac{m}{\sigma}}{2r} h' = 0. \qquad (3.12)$$

(3.10) is compatible with this provided that

$$\sigma = \frac{m}{3m+2q},$$

which is the frequency-wavenumber relationship for the prescribed topographic profile. *Table 3.2* lists the frequencies for a sequence of topography parameters q and the wavenumbers m = 1,2. The table indicates that *the topography has a dominant effect on the periods.* The solutions

$$\psi(r) = Ar^m (1-r^q), \quad \sigma = \frac{m}{3m+2q}$$

embrace all those motions whose stream function ψ has no radial nodal circle. Hence, they contain in particular the solutions for the parabolic depth profile as shown in the left and right columns of *Figure 3.1*.

q	m = 1	m = 2
0.5	0.250	0.286
1	0.200	0.250
2	0.143	0.200
5	0.0769	0.125

Table 3.2

Topography effect on the dimensionless eigenfrequency of the two first modes in the model of Saylor et al. (1980).

Let us see how the above solutions have been used to explain observations:

(1) *Figure 1.2b* indicates that the bathymetric profile for Southern Lake Michigan can reasonably be approximated by (3.10) if q = 1+ε, where ε is small. With ε = 0.25 the period becomes T = 96 hours which coincides with the observed period of the TW in Southern Lake Michigan.

(2) The computational results of the wind driven currents in the Bornholm basin as obtained by Simons (*Figure 1.16*) suggest that this system of gyres might be interpretable as a TW. Wenzel (1978) inferred from this figure and Simons' computations a period between 11 and 14 days. Comparison of *Figures 1.16 and 3.1* shows that a mode with (n, m) = (0, 2) might approximate the flow pattern determined by Simons. With sufficient tolerance for interpretation this is so, but the period as obtained with a parabolic profile (which approximates the topography reasonably) is only 4-5 days. Thus Wenzel suggested that the flow configuration in the Bornholm basin might be the interior part of a mode with one or more nodal circles. With (n, m) = (1, 2) the period is 8 days (see *Table 3.1*), but (n,m) = (2,2) yields T = 13.8 days. This mode was numerically computed by fitting a 10th degree polynomial (the period of this solution is 12.6 days) to the mean topography of the Bornholm basin and solving the two-point boundary value problem (3.6) numerically. The radial section of the surface elevation of this solution is shown in *Figure 3.2*. Wenzel argues that the system of gyres in the outer two rings (r ≥ 0.8) should be discarded when comparing this theoretical solution with Simons' *Figure 1.16* since in the real situation the Bornholm island disturbs the boundary behavior considerably. We personally think that this interpretation stretches the potentials of this theoretical model beyond reasonable tolerance. This particular configuration may require a model which accounts for an island, see section 3.4.

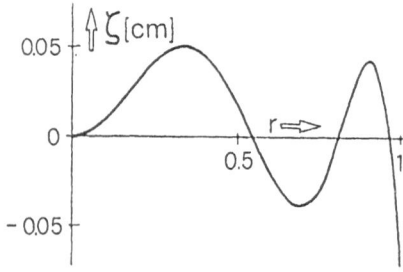

Figure 3.2

Radial cross-section of the surface elevation of the (2, 2)- mode in a circular basin with a depth-profile represented by a polynomial of degree 10.
[From Wenzel, 1978]

3.3 Elliptic basin with parabolic bottom

Most lakes are long in one direction and not well approximated by circles. It is interesting, therefore, to see how the periods and mode structures of TW's depend on the aspect ratio (i.e. the width to length ratio) of the basins.

We consider the TW-equation in dimensionless form and in Cartesian coordinates

$$(h^{-1} \Psi_{xt})_x + (h^{-1} \Psi_{yt})_y - h_x^{-1} \Psi_y + h_y^{-1} \Psi_x = 0, \quad \text{in } \mathcal{D},$$

$$\Psi = 0, \quad \text{on } \partial\mathcal{D}, \tag{3.13}$$

and choose a parabolic depth profile

$$h = \frac{1}{2}((1-a)x^2 + (1+a)y^2) - 1, \tag{3.14}$$

where $A(h) = \sqrt{2(h+1)/(1-a)}$ and $B(h) = \sqrt{2(h+1)/(1+a)}$ are the semi-axes of the elliptic depth-contours. These have all identical aspect ratios

$$r = \frac{B(h)}{A(h)} = \sqrt{\frac{1-a}{1+a}}, \quad a = \frac{1-r^2}{1+r^2}, \tag{3.15}$$

and the profile has a maximum depth $|h|_{max} = 1$. The basin is bounded by the zero depth contour line, an ellipse with $A(0)$ and $B(0)$ as semi-axes.

The following analysis is due to Ball (1965). With (3.14) and the transformation

$$\psi = h^{-2} \Psi$$

it is straightforward to show that (3.13) takes the form

$$4\psi_t + 3((1-a)x \, \psi_{xt} + (1+a)y \, \psi_{yt}) + h(\psi_{xxt} + \psi_{yyt}) +$$

$$+ (1-a)x \, \psi_y - (1+a)y \, \psi_x = 0, \quad \text{for } h < 0, \tag{3.16}$$

$$\psi = \text{finite}, \quad \text{for } h = 0.$$

Note that the boundary condition $\Psi = 0$ along $\partial\mathcal{D}$ necessarily requires that ψ is bounded on $\partial\mathcal{D}$. The velocities are given by

$$u = -h^{-1}(h^2 \psi)_y = -2h_y \psi - h\psi_y,$$

$$v = h^{-1}(h^2 \psi)_x = 2h_x \psi + h\psi_x.$$

The advantage of the introduction of the stream function ψ is that (3.13) transforms into a differential equation with the following special property. Suppose, ψ is an even (odd) polynomial of degree N, then the differential equation (3.16) generates again an even (odd) polynomial of the same degree.

Taking advantage of this fact, we consider first a polynomial with degree $N = 0$, i.e. a constant ψ_{00} which obviously satisfies (3.16). This is a *simple steady gyre* with the velocity field

$$(u, v) = 2\psi_{00} (-(1+a)y, (1-a)x),$$

representing an elliptical rotation with constant vorticity $4\psi_{00}$. Maximal speeds are experienced along the shore-line.

More insight provides the choice of a homogeneous odd polynomial of degree $N = 1$, the *linear Ball-mode*:

$$\psi_1 = \psi_{10}(t) \cdot x + \psi_{01}(t) \cdot y. \tag{3.17}$$

Substitution into (3.16) yields the coupled system

$$(7-3a) \dot{\psi}_{10} + (1-a) \psi_{01} = 0 , \tag{3.18}$$
$$(7+3a) \dot{\psi}_{01} - (1+a) \psi_{10} = 0 ,$$

with $(\ )^{\cdot} \equiv d/dt$. Assuming a harmonic time evolution $e^{-i\sigma t}$ for both coefficient functions, (3.18) allows nontrivial ψ_{10} and ψ_{01} if and only if

$$\sigma^2 = \frac{1-a^2}{49-9a^2} . \tag{3.19}$$

This relation describes the dependence of the frequency on the aspect ratio parameter r (via a, see (3.15)). *Table 3.3* lists the periods obtained with (3.19). Obviously, $a = 0$ recovers the solution for the circle with parabolic bottom profile. Smaller a results in smaller σ; consequently, the more elongated the ellipses become the larger will be the periods. In view of observational results for Lakes of Lugano and Zurich, reported in Chapter 1 this is unfortunate as these lakes are long and narrow, and measurements point at oscillations with periods of 3-4 days. This is lower than the 118 hours obtained as a *lower* bound for the fundamental linear Ball mode.

The linearity of (3.17) implies that the line $\psi_1 = 0$ which separates

r	a	linear		quadratic	
		σ	T [h]	σ	T [h]
1.0	0	0.143	118	0.200	84.5
0.67	0.385	0.134	126	0.190	88.8
0.50	0.600	0.118	143	0.173	97.7
0.33	0.800	0.091	185	0.139	121
0.1	0.980	0.031	542	0.051	335
0	1	0	∞	0	∞

Table 3.3

Frequencies and periods of the linear and quadratic Ball-mode for various aspect ratios r. The periods are calculated with $f = 2\pi/16.9$ h.

vortices of different signs is a straight line which, owing to (3.18), rotates anticlockwise around the basin. *Figure 3.3a* shows the time evolution of this mode. The structure of the wave pattern i.e. the number of gyres is conserved in the course of a wave cycle. This is in accord with the wave patterns found in the circular basin.

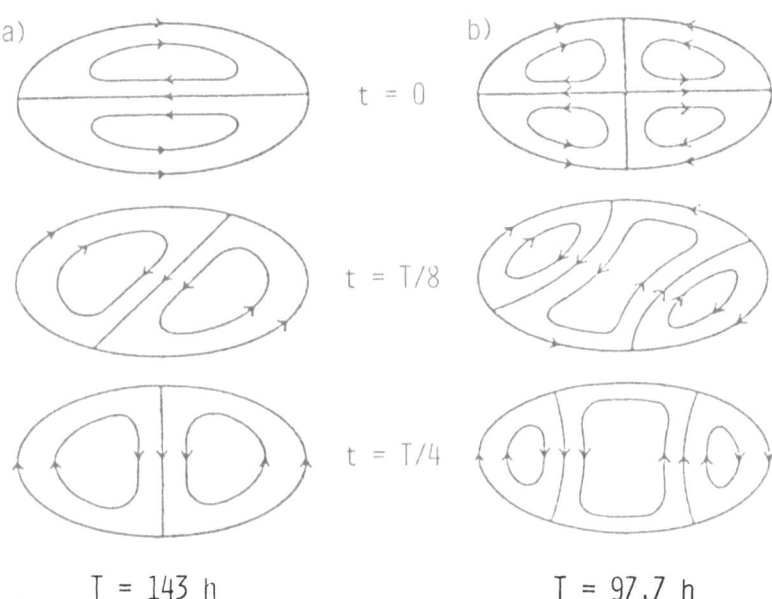

a)   t = 0   b)

t = T/8

t = T/4

T = 143 h   T = 97.7 h

Figure 3.3

Mass transport stream line patterns for the "linear" (a) and "quadratic" (b) mode of the TW-equation in an elliptic basin with parabolic bottom profile. [From Ball, 1965]

To obtain the next higher mode, we select an even polynomial of degree $N = 2$.

$$\psi_2 = \psi_{00} + \psi_{20}\, x^2 + \psi_{11}\, xy + \psi_{02} y^2, \qquad (3.20)$$

with the time dependent coefficient functions $\psi_{mn}(t)$. (3.20) characterizes the *quadratic Ball-mode*. Substitution into (3.16) and equating equal powers of x and y, respectively, yields the system

$$(11-7a)\,\dot{\psi}_{20} + (1-a)\,\dot{\psi}_{02} + 2(1-a)\,\psi_{11} = 0,$$
$$10\,\dot{\psi}_{11} + (1-a)\,\psi_{02} - (1+a)\,\psi_{20} = 0,$$
$$(11+7a)\,\dot{\psi}_{02} - (1+a)\,\dot{\psi}_{20} - 2(1+a)\,\psi_{11} = 0, \qquad (3.21)$$
$$2\dot{\psi}_{00} - \dot{\psi}_{20} - \dot{\psi}_{02} = 0,$$

which allows periodic solutions proportional to $e^{-i\sigma t}$ provided that

$$\sigma(5\sigma^2(5-2a^2)-(1-a^2)) = 0.$$

Again, there is a steady solution $\sigma = 0$ and an oscillating solution with

$$\sigma^2 = \frac{1-a^2}{5(5-2a^2)} . \qquad (3.22)$$

Table 3.3 collects frequencies and periods for several aspect ratios r. For a fixed aspect ratio the periods of the quadratic mode are smaller than those of the fundamental linear mode.

As (3.21) indicates, a steady solution must have

$$\psi_{11}^{st} = 0 \quad \text{and} \quad \psi_{02}^{st} = \frac{1+a}{1-a} \psi_{20}^{st} ,$$

and hence

$$\psi_2^{st} = \psi_{00} + A ((1-a) x^2 + (1+a) y^2) ,$$

where ψ_{00} and A are constants. $\psi_{00} \neq 0$, $A = 0$ recovers the simple steady gyre whereas $\psi_{00} = 0$, $A \neq 0$ yields the steady second order solution

$$\psi_2^{st} = h^2 \psi_2^{st} = 2 Ah^2 (h+1) \qquad (3.23)$$

This stream function vanishes along the boundary ($h = 0$) and at the center $(x, y) = (0, 0)$ and is positive otherwise; furthermore, its value is constant along similar ellipses and assumes a maximum value along the ellipse with $h = -2/3$ between the center and the shore line, *Figure 3.4a.* The steady flow corresponding to the solution (3.23) is qualitatively indicated in *Figure 3.4b.* An anticyclonic elliptical gyre in the center is surrounded by an elliptical ring of cyclonically rotating fluid.

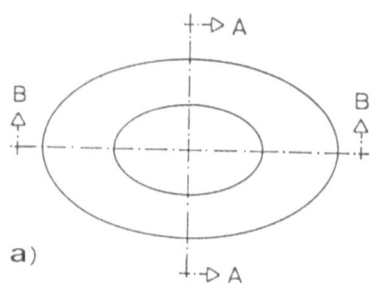

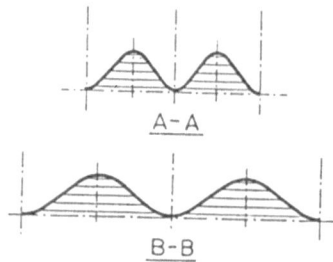

a)

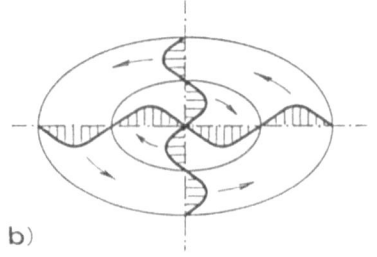

b)

Figure 3.4

a) Distribution of the mass transport stream function and

b) of the associated velocity field of the "quadratic" steady solution (3.23).

Oscillating solutions are obtained by constructing the eigenvector of (3.21) corresponding to the frequency given by (3.22). We quote Ball's result (real.part)

$$\psi_2 = h^2 \psi_2 = \Lambda h^2 \left[((1+a)(3-2a) y^2 - (1-a)(3+2a) x^2 + a) \sin \sigma t \right.$$
$$\left. + (6(1-a^2)/5\sigma) xy \cos \sigma t \right]. \tag{3.24}$$

For $t = 0$ the nodal lines $\psi_2 = 0$ are the lines $x = 0$ and $y = 0$, whereas for $t > 0$ they are rotating hyperbolas (note that $0 < a < 1$). As illustrated in *Figure 3.3b* the wave pattern starts with four gyres of which the two positive vortices merge together building three gyres in the basin for most part within a period. The structure of this mode is therefore *not conserved* during the cycle. This is a new phenomenon due to the influence of the aspect ratio parameter a.

3.4 Elliptic basin with exponential bottom

a) Basin with central island

In the previous sections TW's in circular and specific elliptical domains were discussed. Whereas the model of Saylor et al. uncovered a conspicuous dependence of the frequencies on the topography, Ball's model enabled investigation of the effect of the aspect ratio. In this section we present models which account for both bathymetric parameters and therefore permit a more realistic modelling of the lake basin.

To introduce a topography parameter in an elliptical basin, Mysak (1985) set out to study the TW-equation in elliptical coordinates (ξ, η). Basically, this was a generalization of Saylor's choice who studied a circular domain in polar coordinates and thus lost the possibility of incorporating into the analysis an aspect ratio parameter.

The derivation of the TW-equation in the elliptic coordinate system has already been given in section 2.5; the result was

$$(h^{-1} \psi_{\xi t})_\xi + (h^{-1} \psi_{\eta t})_\eta - h_\xi^{-1} \psi_\eta = 0, \qquad \begin{matrix} 0 < \xi < \xi_S, \\ 0 \le \eta \le 2\pi, \end{matrix} \tag{3.25}$$
$$\psi = 0, \qquad \xi = \xi_S, \ 0 \le \eta \le 2\pi,$$

where ξ is the radial and η the azimuthal coordinate, ξ_S is the elliptic shore-line. Note that ξ_S is related to the aspect ratio parameter r through

$$r = \frac{B}{A} = \frac{a \sinh \xi_S}{a \cosh \xi_S} = \tanh \xi_S. \tag{3.26}$$

The velocity field can easily be computed from the stream function ψ by means of the formulas

$$u_\xi = -(hJ)^{-1} \psi_\eta, \qquad u_\eta = (hJ)^{-1} \psi_\xi, \qquad (3.27)$$

and the definitions of a and the Jacobian J are listed in *Table 2.3*.

Consider now a topography which is constant along lines of constant ξ (confocal ellipses), and hence has $h_\eta = 0$. For this case (3.25) is a differential equation with constant coefficients provided

$$h_\xi/h = \text{const.}$$

Therefore, we select an exponential depth-profile (shelf) of the form $h(\xi) = \exp(-b\xi)$, $b > 0$. Introducing the separation of variables solution

$$\psi(\xi,\eta) = F(\xi)\, e^{i(m\eta - \sigma t)}, \qquad (3.28)$$

with integer $m > 0$, (2π-periodicity in η), (3.25) becomes

$$F'' + bF' + (\frac{mb}{\sigma} - m^2)F = 0. \qquad (3.29)$$

This second order ODE requires two boundary conditions. No mass transport through the basin boundary ξ_S leads to

$$F(\xi_S) = 0. \qquad (3.30a)$$

To obtain the second boundary condition we consider an elliptic basin with a *central island* in the domain $0 \leq \xi \leq \xi_I$. Hence, an additional no-flux condition must hold at ξ_I,

$$F(\xi_I) = 0. \qquad (3.30b)$$

Mysak's choice was the limit of a barrier-like island $\xi_I = 0$; here ξ_I is retained as a further bathymetric variable. (3.29) has the general solution

$$F(\xi) = e^{-b\xi/2} (A \sin \lambda\xi + B \cos \lambda\xi), \qquad (3.31)$$

with $\lambda^2 = mb/\sigma - m^2 - b^2/4$, and the trigonometric functions imply the frequency of the m-th azimuthal mode to be bounded by

$$0 < \sigma_m < \frac{mb}{m^2 + \frac{b^2}{4}}. \qquad (3.32)$$

The lower bound can be obtained by multiplying (3.29) with F and integrating the resulting equation from $\xi = \xi_I$ to $\xi = \xi_S$. This yields the Rayleigh quotient

$$\frac{mb}{\sigma} = -\frac{\int_{\xi_I}^{\xi_S} (F'' F - m^2 F^2)\, d\xi + \frac{b}{2} \int_{\xi_I}^{\xi_S} \frac{d}{d\xi}(F^2)\, d\xi}{\int_{\xi_I}^{\xi_S} F^2\, d\xi} = \frac{\int_{\xi_I}^{\xi_S} (F'^2 + m^2 F^2)\, d\xi}{\int_{\xi_I}^{\xi_S} F^2\, d\xi} > 0.$$

The boundary conditions (3.30) discretize the radial wavenumber λ such that $\sin \lambda(\xi_S - \xi_I) = 0$ and hence

$$\lambda = \frac{n\pi}{\xi_S - \xi_I}, \quad n = 1, 2, \ldots .$$

With this and the definition of λ we obtain the eigenfrequency as a function of the three bathymetric parameters ξ_S, ξ_I, b

$$\sigma = \frac{mb}{m^2 - \frac{b^2}{4} + (\frac{n\pi}{\xi_S - \xi_I})^2} . \tag{3.33}$$

Table 3.4 demonstrates the effect of aspect ratio and island parameter for the three modes $(m,n) = (1,1)$, $(2,1)$ and $(2,2)$, while the finite shore-line depth $h(\xi_S) = 0.1$ and the depth at the island $h(\xi_I) = 1$ are kept constant. $\xi_I/\xi_S = 0$ recovers the case studied by Mysak (1985) and it is seen from (3.33) that the eigenfrequencies for the barrier-island $\xi_I = 0$ are an upper bound. We recognize that the influence of the aspect ratio on

r	ξ_S	ξ_I/ξ_S	b	σ		
				(1,1)	(2,1)	(2,2)
0.99	2.65	1/2	1.74	0.235	0.335	0.127
		1/3	1.31	0.284	0.344	0.153
		0	0.87	0.335	0.311	0.177
0.67	0.81	1/2	5.72	0.082	0.156	0.045
		1/3	4.29	0.108	0.200	0.059
		0	2.86	0.156	0.269	0.085
0.5	0.55	1/2	8.38	0.056	0.110	0.031
		1/3	6.29	0.074	0.144	0.041
		0	4.19	0.110	0.204	0.060
0.33	0.35	1/2	13.3	0.036	0.071	0.019
		1/3	9.97	0.047	0.093	0.026
		0	6.64	0.071	0.137	0.039
0.1	0.10	1/2	45.9	0.010	0.021	0.006
		1/3	34.4	0.014	0.027	0.008
		0	22.9	0.021	0.041	0.011

Table 3.4

Influence of the aspect ratio and relative island width on the eigenfrequencies.

the eigenfrequency is stronger than that of the topography. *Figure 3.5* displays the contour lines of the stream functions ψ_{11} and ψ_{21}. These correspond to the linear and quadratic Ball modes, respectively, shown in *Figure 3.4*. The effect of the barrier island is obvious: during one cycle the gyres have to pass the narrow gap between $\xi_I = 0$ and ξ_S at the foci. There, the contour lines are accumulated and large gradients of ψ are observed. Hence, these points will be critical with respect to the velocity field. Inserting the real part of (3.28) into (3.27) yields

$$u_\xi = \frac{1}{J} m e^{b\xi/2} \sin \lambda (\xi - \xi_I) \sin (m\eta - \sigma t),$$

$$u_\eta = -\frac{1}{J} \cdot e^{b\xi/2} \left[\lambda \cos \lambda (\xi - \xi_I) - \frac{b}{2} \sin \lambda (\xi - \xi_I) \right] \cos (m\eta - \sigma t),$$

with $J = a\sqrt{\sinh^2\xi + \sin^2\eta}$ and $\lambda = n\pi/(\xi_S - \xi_I)$. Indeed, the limits $(\xi_I, \eta) \to (\xi_I, \pi)$ or $(\xi_I, 0)$ of u_η do not exist for $\xi_I \to 0$. Although the case $\xi_I = 0$ is far from a realistic situation it can serve as a reasonable estimate for the eigenfrequencies for a basin with no island. In the next subsection we discuss an analytical model for this configuration.

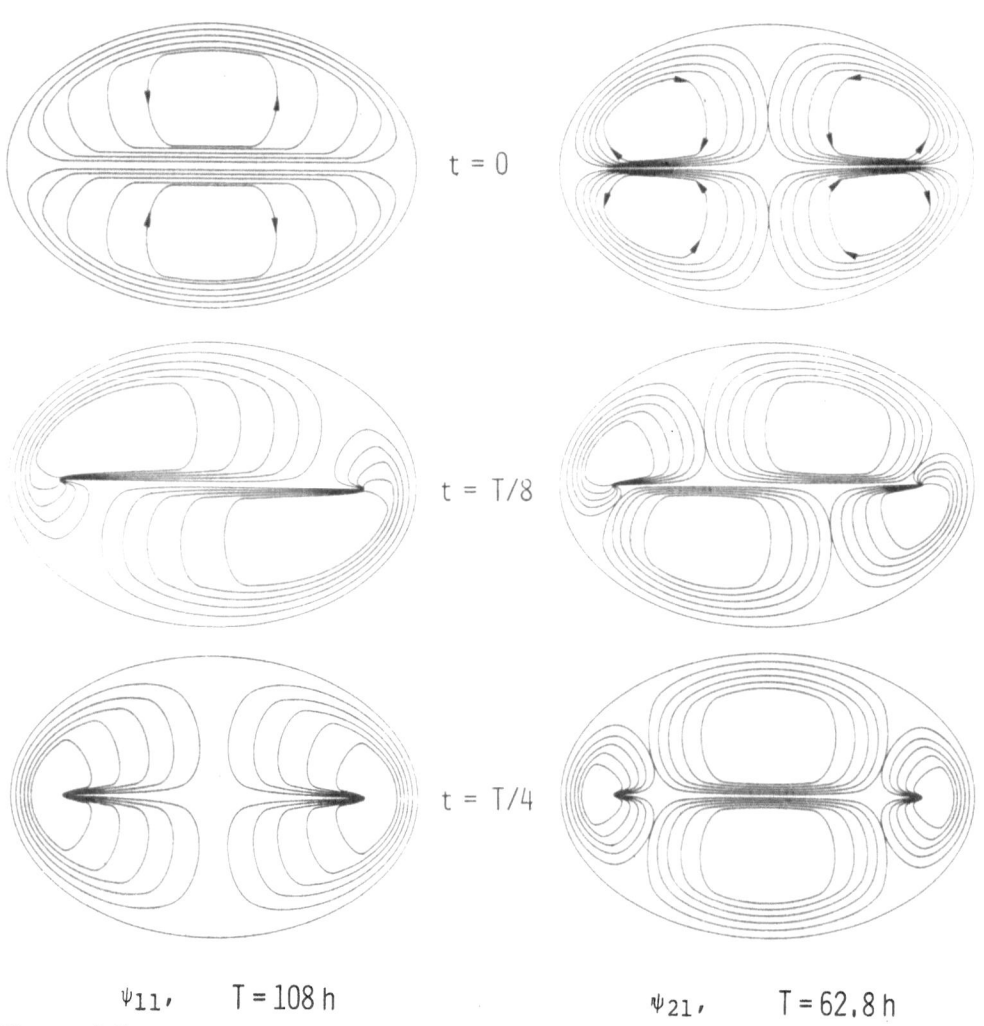

$\psi_{11}, \quad T = 108\,h$ $\qquad\qquad$ $\psi_{21}, \quad T = 62,8\,h$

Figure 3.5

Contours of the stream functions ψ_{11} and ψ_{21} for an elliptic basin with a barrier. The parameters are $\xi_S = 0.805$, $\xi_I = 0$, $b = 2.86$. [From Mysak, 1985]

Let us see, how the above solutions can be used to interprete observations made in Lake of Zurich (Mysak, 1985). Figure 1.11 explains the bathymetry of the Lake of Zurich and the mooring positions of the 1978 summer study carried out by our limnology group. Figure 1.12 exhibits spectra of the isotherm depth time series taken at the indicated moorings. These spectra disclose a persistent peak at approximately 100-110 hours, and it is demonstrated that the phase of the corresponding wave propagates in the anticlockwise direction around the basin. Finally, Figure 1.13 indicates that the velocity vector at the 100-110 hour period rotates counterclockwise (clockwise) at the middle (near shore) stations 6 and 7. All these properties fit the elliptical topographic wave model derived above.

Figure 3.6 shows a depth profile along the thalweg (solid line) and a model fit (dashed line) of the Northern basin of the Lake of Zurich. From this figure we infer that suitable values for D_S and D are 37 m and 130 m, respectively which imply $h(\xi_S) = D_S/D = 0.29$. From *Figure 1.11* it appears that a reasonable choice for r (the aspect ratio) is perhaps 0.1, which implies $\xi_S = 0.1$ (if $\xi_I = 0$). Such a small value of ξ_S together with the above value of $h(\xi_S)$, however, yields a very large value of the period (several

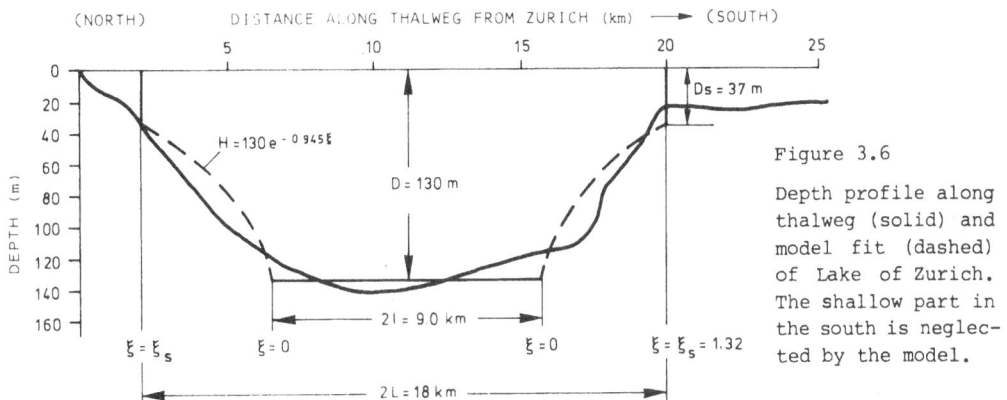

Figure 3.6

Depth profile along thalweg (solid) and model fit (dashed) of Lake of Zurich. The shallow part in the south is neglected by the model.

weeks). Moreover, for a lake with $\xi_S = 0.1$, all the contours at each end of the lake are crowded together in a very small region, which is not at all the case for Lake of Zurich in *Figure 1.11*. To obtain a more faithful representation of the contours at the lake ends, a large value of ξ_S is required. Mysak proposes to select values of ℓ and L from the thalweg profile such that this profile is "best represented" (see *Figure 3.6*). With $a = \ell/L = 0.5$ and $h(\xi_S) = 0.288$ we obtain $\xi_S = 1.32$ and $b = 0.945$. The aspect ratio of this model lake takes the value $r = \tanh \xi_S = 0.866$, corresponding to a lake which is much wider than in reality. Inserting these values in (3.33) yields for the fundamental mode (with $\xi_I = 0$) a period $T = 124$ hours. This is reasonably close to 110 h. However, it may be objected that the large aspect ratio is not representative for this lake. This is true, but the procedure provides a first hint that the topography at the lake ends and the modelling of the depth-lines in that region are physically more important than a true representation of the aspect ratio and along with it a satisfactory fit of the long sides of the elongated lake. We will focus our attention on these facts in Chapters 6 and 7.

b) *Basin without island*

The analysis which is outlined here extends the theory of subsection a) and is due to Mysak et al. (1985) with corrections by Johnson (1986).

It is characteristic of the elliptical coordinate system that the formulation of the boundary condition at the center $\xi = 0$ is subtle. It is necessary to have both ψ and $\nabla\psi$ continuous "across" $\xi = 0^{*)}$, in order that the velocity field takes physically meaningful values. Therefore,

---

*) *Clearly, in elliptical coordinates* $\xi \geq = 0$. *Continuity of a quantity* $\phi(\xi,\eta)$ "across" $\xi = 0$ *means* $\lim_{\xi \downarrow 0} \phi(\xi, 2\pi-\eta) = \lim_{\xi \downarrow 0} \phi(\xi,\eta)$, $0 < \eta < 2\pi$.

ansatz (3.28) is too restrictive to fulfil the extended boundary condition at the center. The trial solution (3.28) is complemented by the
contribution with negative integers $m < 0$. Thus following Johnson (1986),
we write

$$\psi(\xi,\eta) = F_1(\xi)\, e^{i(m\eta-\sigma t)} + F_2(\xi)\, e^{i(-m\eta-\sigma t)}, \quad m = 1,2,\ldots . \tag{3.34}$$

Again using the shelf profile $h(\xi) = \exp(-b\xi)$ equation (3.25) is equivalent
to the system

$$F_1'' + b\, F_1' + \left(\frac{mb}{\sigma} - m^2\right) F_1 = 0,$$
$$F_2'' + b\, F_2' + \left(-\frac{mb}{\sigma} - m^2\right) F_2 = 0, \tag{3.35}$$

with $(\ )' = d/d\xi$ and the four boundary conditions

$$F_1(\xi_S) = 0, \qquad F_2(\xi_S) = 0, \tag{3.36a}$$

$$F_1(0) - F_2(0) = 0, \tag{3.36b}$$

$$F_1'(0) + F_2'(0) = 0. \tag{3.36c}$$

System (3.35) together with (3.36) is a well-posed boundary value problem of second order in the interval $[0,\xi_S]$, which can be solved in
terms of exponential functions. The conditions (3.36) will select the
eigenfrequencies.

Because of the form of (3.35) and (3.36) the radial functions can be
taken as purely real, and it can be verified that

$$F_1(\xi) = e^{-\frac{b}{2}\xi} \sin \lambda_1(\xi_S-\xi), \qquad F_2(\xi) = e^{-\frac{b}{2}\xi} \sinh \lambda_2(\xi_S-\xi),$$
$$\lambda_1^2 = \frac{bm}{\sigma} - m^2 - \frac{1}{4}b^2, \qquad \lambda_2^2 = \frac{bm}{\sigma} + m^2 + \frac{1}{4}b^2, \tag{3.37}$$

fulfil (3.35) and (3.36a,b). (3.36c) eventually requires

$$\lambda_1 \cot \lambda_1 \xi_S + \lambda_2 \coth \lambda_2 \xi_S + b = 0, \tag{3.38}$$

from which the eigenfrequencies can be calculated. Note that for sufficiently large m and σ λ_1^2 in (3.37) becomes negative and F_1 takes the
same form as F_2. The cot in (3.38) equally transforms into a coth, and
then real eigenfrequencies are no longer allowed; σ is thus bounded according to (3.32). Equation (3.38) yields a countable set of eigenfrequencies for given topography parameter b and azimuthal wavenumber m
and for each σ the inequalities

$$(n-\frac{1}{2})\pi < \lambda_1(\sigma)\cdot\xi_S < n\pi, \quad n = 1,2,3,\ldots$$

must hold. Table 3.5 gives eigenfrequencies calculated by Johnson (1986)
for $\xi_S = 0.805$ (ellipse with aspect ratio $r = 2/3$) and $b = 2.86$ (shore line
depth $h(\xi_S) = 0.1$). Figure 3.7 displays the stream line contours of the

m	n = 1	2	3	4
1	0.201	0.0541	0.0235	0.0130
2	0.327	0.102	0.0458	0.0257
3	0.376	0.139	0.0659	0.0375
4	0.379	0.165	0.0830	0.0483

Table 3.5

Eigenfrequencies of the first TW-modes in an elliptic basin. The parameters are $\xi_S = 0.805$ and $b = 2.86$.

r	$h(\xi_S) = 0.1$	0.2	0.5	0.8
0.99	0.380	0.281	0.128	0.0422
0.67	0.201	0.158	0.0783	0.0270
0.50	0.143	0.113	0.0569	0.0197
0.33	0.0921	0.0732	0.0371	0.0129
0.1	0.0270	0.0216	0.0110	0.00384

Table 3.6

Topography and aspect-ratio effect on the frequency of the (1,1)-mode.

modes with $(m,n) = (1,1)$, $(2,1)$ and $(1,2)$. The patterns resemble those of Ball's model or Mysak's island model and modifications here are due to the different choice of the topography (with respect to Ball (1965)) and of the central boundary condition (with respect to Mysak (1985)).

The influence on the fundamental mode of the variation of both bathymetric parameters ξ_S (via aspect ratio) and b (via shore line depth) is shown in Table 3.6. The influence due to topography is dominant.

Table 3.7 substantiates the fact that the eigenfrequencies obtai-

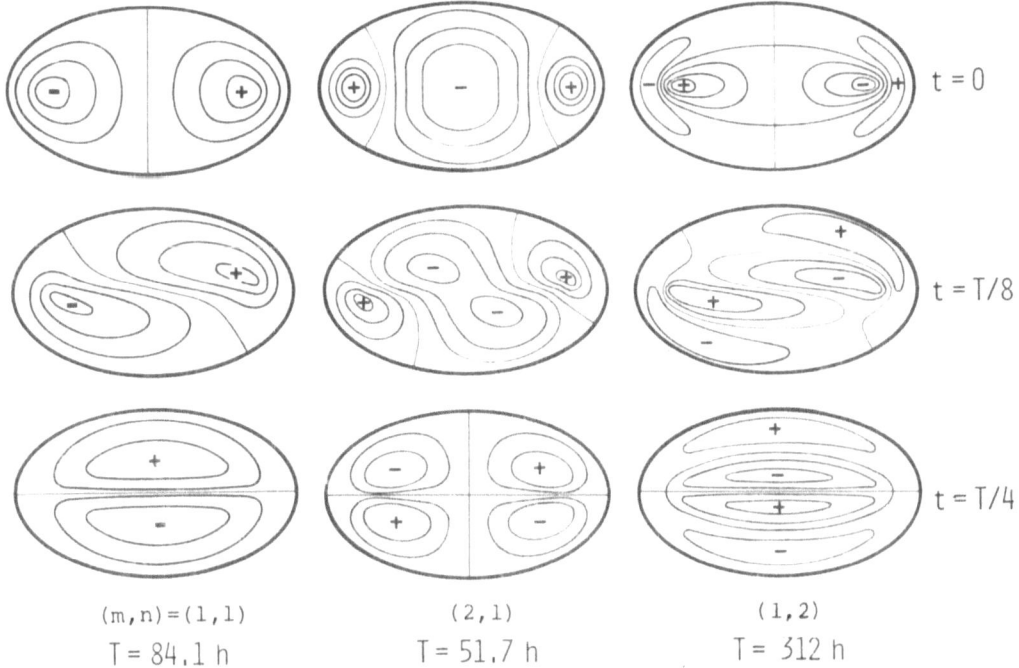

$(m,n) = (1,1)$	$(2,1)$	$(1,2)$
$T = 84.1$ h	$T = 51.7$ h	$T = 312$ h

Figure 3.7 Stream line contours of the three lowest modes in an elliptical lake with exponential bottom. [From Johnson, 1986]

ned with equation (3.33) for $\xi_I = 0$ are estimates for those in the simple basin, and it is seen that the estimate is better for large aspect ratios and small shore line depths.

Johnson extends his model also towards more realistic bottom profiles. The purely exponential profile $h(\xi) = e^{-b\xi}$ exhibits an unrealistic topography in the neighbourhood of $\xi = 0$ in that the basin has the form of a trench there. He thus investigates a profile given by

$$
h(\xi) = \begin{cases} e^{-b(\xi-\xi_B)}, & \xi_B \leq \xi \leq \xi_S, \\ 1, & 0 \leq \xi \leq \xi_B, \end{cases} \tag{3.39}
$$

r	$h(\xi_S) = 0.1$		0.5	
	$\sigma_{barrier}$	σ	$\sigma_{barrier}$	σ
0.99	0.335	0.380	0.108	0.128
0.67	0.156	0.201	0.052	0.078
0.50	0.110	0.143	0.037	0.057
0.33	0.071	0.092	0.024	0.037
0.1	0.021	0.027	0.007	0.011

Table 3.7
Comparison of the eigenfrequencies of the first TW-mode in a basin with a barrier and a basin without island.

m	n = 1	2	3	4
1	0.137	0.0321	0.0129	0.0068
2	0.253	0.0629	0.0256	0.0135
3	0.340	0.0916	0.0379	0.0202
4	0.397	0.117	0.0496	0.0266

Table 3.8
Eigenfrequencies of TW's in an elliptic basin with flat bottom. The parameters are $\xi_S = 0.805$, $\xi_B = 0.434$, $b = 6.2$.

describing a basin with flat bottom in its center. Trials of a slightly modified form of (3.34) can be formulated for the domains $\xi < \xi_B$ and $\xi > \xi_B$, respectively. A matching condition requiring continuity of ψ and ψ_ξ must be satisfied at $\xi = \xi_B$. We do not intend to go into any details of this calculation since it can be performed given the experience of the previous derivations. Table 3.8 lists the frequencies for the parameters $\xi_B = 0.805$ (aspect ratio $r = 2/3$), $\xi_B = 0.434$ (flat bottom occupying half the basin width) and $b = 6.2$ (shore line depth $h(\xi_S) = 0.1$).

Comparison of Tables 3.5 and 3.8 indicates that increasing the width of any flat region whilst holding shore line depth and aspect ratio of the basin fixed decreases the eigenfrequencies. Investigation of the stream function (not presented here) shows that largest speeds occur in the domain $\xi_B \leq \xi \leq \xi_S$ and that the stream lines covering the flat part of the lake are nearly straight lines.

So far we have studied the TW-equation in various coordinate systems for closed basins. i.e. finite domains. It was demonstrated that the discrete *spectrum* exhibits conspicuous dependencies on the bathymetric parameters, such as the width to length ratio (aspect ratio) and the topography parameter. Generally, increasing the topographic gradients, i.e. $|h'/h|$, decreases frequencies considerably. The same but weaker influence is experienced when the aspect ratio is decreased.

In the next section we intend to briefly present solutions of the vorticity equation in simpler configurations. These are domains which are infinite or semi-infinite with respect to one or both coordinates. These *infinite domains* have a *continuous spectrum* and the properties of it will be studied extensively in further sections.

3.5 Topographic waves in infinite domains

Major developments in the understanding of second class waves were not advanced by solving the TW-equation in *finite* domains, but rather by studying these waves in *infinite* domains, such as channels, continental shelves, trenches, etc. Our aim is not to provide a complete account of the history and the availability of known solutions, for this the reader is directed to Mysak (1980a,b) and more recent literature[*]. By presenting solutions to analytically accessible configurations we would rather like to work out the typical physical properties of the characteristics of these guided waves. The results presented here will be preparatory to developments in subsequent chapters (particularly Chapter 5).

In the relevant literature, TW's are usually studied in the context of shallow water waves in a homogeneous or stratified fluid medium on the f- or β-plane. Much effort is directed towards an understanding of their generation, propagation and their modification by topography, stratification, nonlinearities etc. Motivation are observations in the continental shelf regions of the Earth's oceans, and these observations have often found excellent theoretical explanations through these models. However, the success in these interpretations was due in parts to the simplicity of the domains and their topographies. This made it possible to study the low-frequency properties of the shallow water equations and to demonstrate existence of continental shelf waves, barotropic and baroclinic Kelvin waves and edge waves and to analyse their interaction

---

[*] *Brink (1980), Brink (1982), Djurfeldt (1984), Gratton (1983), Gratton & LeBlond (1986), Koutitonsky (1985), Lie (1983), Lie & El-Sabh (1983), Ou (1980), Takeda (1984), Mysak et al. (1979).*

(Huthnance, 1975, Allen, 1975). All these interactions will be ignored here, and only the TW-equation (2.22) will be studied in which baroclinic effects and barotropic gravity waves are neglected.

a) Straight channel

Consider a straight, infinite channel with the Cartesian coordinate system as indicated in Figure 3.8. With a depth profile h(y) which is constant along the channel axis and a carrier-wave ansatz of the form

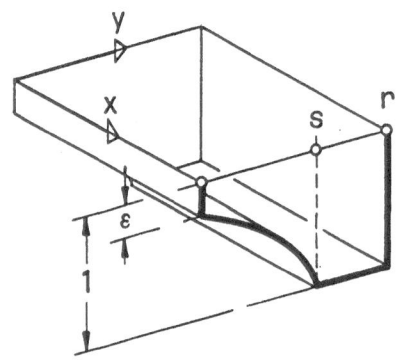

$$\psi = \psi(y)\, e^{i(kx-\sigma t)}, \qquad \sigma = \omega/f$$

the TW-equation reduces to

$$(h^{-1}\,\psi')' - \left[\frac{k}{\sigma}(h^{-1})' + k^2\, h^{-1}\right]\psi = 0, \qquad (3.40)$$

in which $(\ )' = d/dy$.

Equation (3.40) is subject to the familiar no-flux conditions

$$\psi(0) = \psi(r) = 0.$$

Figure 3.8

Infinite channel with one-sided shelf. At the boundary points, 0, s, r the functions h and h' may not be continuous.

Furthermore, at interior points we request that[*]

$$\llbracket \psi \rrbracket = 0,$$
$$\llbracket (\psi' - \frac{k}{\sigma}\psi)/h \rrbracket = 0, \qquad \text{at} \quad y = s. \qquad (3.41)$$

Equation (3.41a) means that the transport is continuous whereas (3.41b) follows by integrating (3.40) across the discontinuity:

$$\lim_{\varepsilon\downarrow 0} \int_{s-\varepsilon}^{s+\varepsilon} (h^{-1}\,\psi')' - \frac{k}{\sigma}(h^{-1})'\,\psi - k^2\, h^{-1}\,\psi)\, dy = 0,$$

$$\lim_{\varepsilon\downarrow 0} \left[\left[h^{-1}\,\psi' - \frac{k}{\sigma}\, h^{-1}\,\psi\right]_{s-\varepsilon}^{s+\varepsilon} + \int_{s-\varepsilon}^{s+\varepsilon}\left[\frac{k}{\sigma}\, h^{-1}\,\psi' - k^2\, h^{-1}\,\psi\right] dy\right] = 0.$$

Because h, ψ and ψ' are all bounded at $y = s$ and h is nonzero at $y = s$ the last integral vanishes in the limit as $\varepsilon \downarrow 0$.

b) Channel with one-sided topography

Consider the piecewise exponential depth profile (see Figure 3.8)

$$h(y) = \begin{cases} \varepsilon e^{by} & 0 \le y \le s, \\ 1 & s \le y \le r, \end{cases}$$

$$b = \frac{1}{s}\ln\frac{1}{\varepsilon}. \qquad (3.42)$$

---

[*] $\llbracket \phi(y) \rrbracket$ at $y = s$ denotes the jump of the quantity ϕ defined by
$$\llbracket \phi(s) \rrbracket = \lim_{\varepsilon\downarrow 0} (\phi(s+\varepsilon) - \phi(s-\varepsilon)).$$

It renders (3.40) an ordinary differential equation with constant coefficients. Subject to the boundary conditions the solution is

$$\psi(y) = \begin{cases} e^{b/2 \cdot y} \sin \lambda y, & 0 \le y \le s, \\ A \sinh k(y-r), & s \le y \le r, \end{cases}$$

$$\lambda^2 = \frac{k}{\sigma} b - k^2 - \frac{b^2}{4} .$$

Evaluating (3.41) yields the *implicit dispersion relation*

$$\frac{1}{\lambda} \tan s\lambda = \frac{-1}{k \coth k(r-s) + \frac{b}{2}} , \qquad (3.43a)$$

$$\lambda^2 = \frac{k}{\sigma} b - k^2 - \frac{b^2}{4} , \qquad (3.43b)$$

of TW's in this infinite channel with one-sided topography. *Figure 3.9* displays the lhs and rhs of (3.43a) as functions of λr. The lhs is independent of σ whereas the rhs is a double-valued relation of λ due to (3.43b). For frequencies lower than a critical value there esists a finite number of intersections (σ, λ) or (σ, k). This number increases stepwise with decreasing σ due to the periodicity of the tangent function. Note that λ^2 is only positive provided the signs of k and σ are identical. The dashed curve in *Figure 3.9* shows the lhs for $\lambda^2 < 0$, i.e. the tan is replaced by a tanh. For this case there are no intersections and hence no real pairs (σ, k) satisfying (3.43). This implies that phase propagation is into the positive x-direction for this configuration,

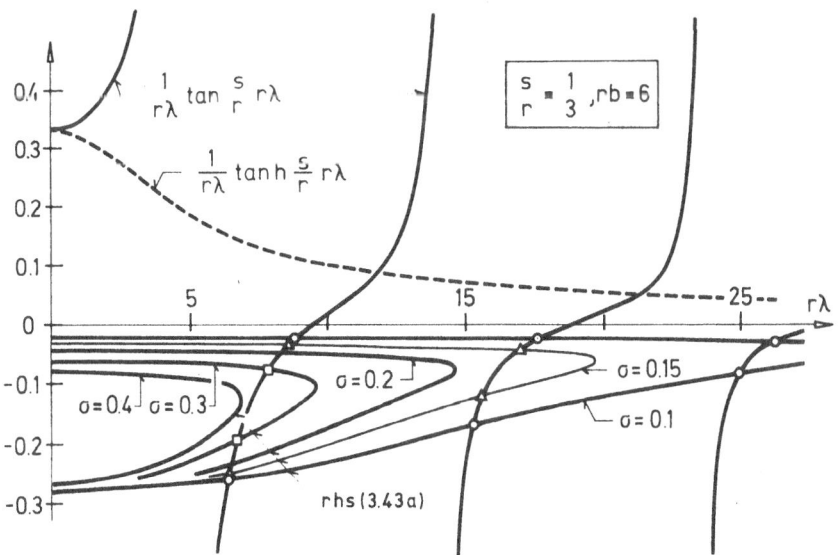

Figure 3.9
Plot of the lhs and rhs of the implicit dispersion relation $\sigma(\lambda)$ or $\sigma(k)$ given in (3.43) for $s/r = 1/3$, $rb = 6$. The points (σ, λ) are indicated with □, △, o.

which amounts to the well-known property of shelf waves on the Northern hemisphere $(f > 0)$: the phase propagation is *right-bounded*[*]. Two limiting cases of this dispersion relation are of interest.

c) *Shelf*

In the limit as $r \to \infty$ the depth profile (3.42) becomes the well-known exponential shelf. Correspondingly the dispersion relation reads

$$\frac{1}{\lambda} \tan s\lambda = - \frac{1}{k + \frac{b}{2}}, \tag{3.44}$$

and the result of Buchwald & Adams (1968) is recovered. *Figure 3.10* displays this dispersion relation for the first five modes. The shapes of these curves exhibit features which are typical of topographycally trapped

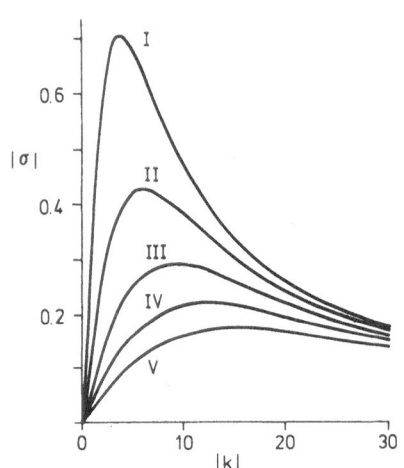

Figure 3.10

Dispersion relation $\sigma(k)$ (3.44) for the first five modes on a shelf with $b = 5.4$.
[From Buchwald & Adams, 1968]

second-class waves: Firstly, long shelf waves are non-dispersive, i.e. as $k \to 0$ $c_{gr} = \partial\sigma/\partial k \to c = \sigma/k > 0$. Phase and group velocity are the same. Secondly, when h'/h[**] is bounded, for $y \in [0,\infty]$ then $c_{gr} < 0$ for some range of $k > 0$. In other words, the dispersion relation $\sigma(k)$ possesses a maximum at $k = k_0$. For $k < k_0$, c and c_{gr} are both positve, and phase and energy propagate in the same direction; for $k > k_0$, c is still positive but c_{gr} is negative. Furthermore, $\sigma(k) \to 0$ as $k \to \infty$. These properties which hold in an infinite domain were proven by Huthnance (1975) in a more general context.

The stream function in the shelf-wave limit takes the form

$$\psi(y) = \begin{cases} e^{b/2 \cdot y} \sin \lambda y, & 0 \leq y \leq s, \\ e^{b/2 \cdot s} \sin \lambda s \cdot e^{-k(y-s)}, & s < y. \end{cases}$$

It decays exponentially for $y > s$ and is essentially sinusoidal in the shelf domain.

---

*) *Right-bounded means that the shallower region is to the right when looking into the direction of phase propagation.*

**) *Because of its significance h'/h is often referred to as slope parameter $S \equiv h'/h$.*

d) *Trench*

A second simple case is obtained when $s \to r$. The dispersion relation is then reduced to

$$\tan r\lambda = 0,$$

whence

$$\sigma = \frac{kb}{k^2 + \left[\frac{b^2}{4} + (\frac{n\pi}{r})^2\right]}, \qquad n = 1,2,\ldots,$$

where the integer n denotes the mode of the wave. The streamfunction of the n-th mode has $n - 1$ nodes across the channel. Long waves in this channel are non-dispersive with phase and group velocity

$$c = c_{gr} = \frac{b}{\frac{b^2}{4} + (\frac{n\pi}{r})^2}, \qquad \text{as} \quad k \to 0.$$

For very short waves the frequency is inversely proportional to the wavenumber, $\sigma = b/k$, and phase and group velocity have opposite sign. The critical point (k_0, σ_0), where the group velocity vanishes is given by

$$(k_0, \sigma_0) = \left[\sqrt{\frac{b^2}{4} + (\frac{n\pi}{r})^2}, \frac{b}{2\sqrt{\frac{b^2}{4} + (\frac{n\pi}{r})^2}}\right].$$

The critical frequency σ_0 strongly depends on the topography parameter $b = h'/h$.

e) *Single-step shelf*

We now demonstrate that boundedness of h'/h is important for the second of Huthnance's properties, namely existence of vanishing group velocity for finite $k_0 < \infty$. To this end consider the profile

$$h(y) = \begin{cases} d, & 0 < y < s, \\ 1, & s < y < r. \end{cases} \tag{3.45}$$

This profile was used by Sezawa and Kanai (1939), Snodgrass et al. (1962) and Larsen (1969) to explain edge waves and trapped long waves. Clearly, since $h' = (1-d)\delta(y-s)$ vanishes everywhere in $y \in (0,r)$ except at $y = s$, where h' is infinite, TW's only exist because of this singularity.

With (3.45) the differential equation (3.40) reduces to $\psi'' - k^2\psi = 0$. Its solutions are

$$\psi(y) = \begin{cases} \sinh ky, & 0 \le y \le s, \\ A \sinh k(y-r), & s \le y \le r. \end{cases}$$

Inserting this into the matching conditions (3.41) readily yields the dispersion relation

$$\sigma = (1-d)\frac{\tanh ks \, \tanh k(r-s)}{d \, \tanh ks + \tanh k(r-s)}. \tag{3.46}$$

Figure 3.11 displays σ(k) for different values of d, and we notice that these are *monotonically growing* functions of k. Indeed

$$\sigma \longrightarrow \frac{1-d}{1+d}, \quad \text{as} \quad k \to \infty,$$

$$\sigma \text{ prop } k, \quad \text{as} \quad k \to 0,$$

and no critical wavenumber k_0 exists. Furthermore, there is only one single fundamental TW mode.

Figure 3.11

Graph of the dispersion relation (3.46) for s/r = 0.3 and 3 values of d. Note σ(rk) is monotone.

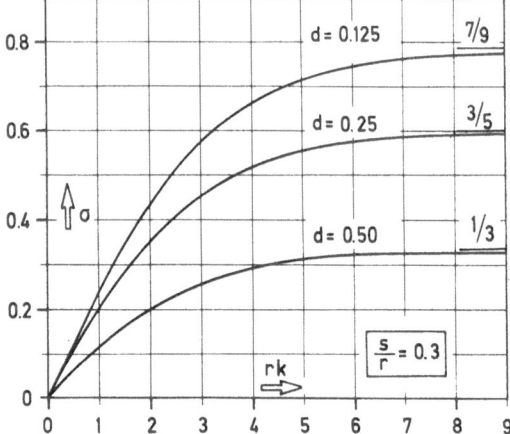

Shelf waves have also been analysed for different topographies with $r \to \infty$. Reid (1958) and Mysak (1968) have investigated the finite width sloping shelf profile of *Figure 3.12a*. The mass transport stream function can in this case be expressed in terms of Laguerre polynomials and dispersion curves are qualitatively as those shown in *Figure 3.11*. However, there are now a countably infinite set of shelf modes because of the sloping portion of the shelf. Ball (1967) on the other hand studied the exponential depth profile of *Figure 3.12b* and finds dispersion curves for shelf waves which are qualitatively as those of *Figure 3.11*, as would be expected. A similar study for escarpment-, trench-, shelf- and wedge waves was given by Djurfeldt (1984).

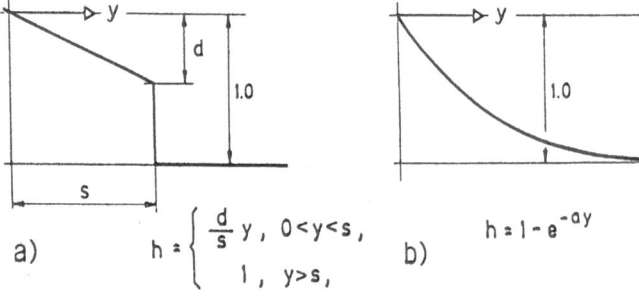

Figure 3.12

a) Finite width sloping shelf profile.

b) Exponential profile.

a) $$h \approx \begin{cases} \dfrac{d}{s}\, y, & 0<y<s, \\ 1, & y>s, \end{cases}$$

b) $h \approx 1 - e^{-ay}$

In reality channels have bathymetric gradients on both sides. From the results obtained so far, it can be concluded that in such a channel (e.g. with a parabolic depth profile) there will be TW's travelling along either sides of the channel each in a right-bounded way. The dispersion relation then consists of two branches (assume σ > 0), one for k > 0 representing those waves trapped to y = 0 and k <0 for those trapped to

$y = r$. This, and the propagation of TW's in *curved* channels will be discussed in Chapter 5.

$\oint$) *Elliptic island*

Island-trapped shelf waves were studied by Mysak (1967), Rhines (1969), Saint-Guily (1972), Buchwald & Melville (1977) and Hogg (1979). All these authors solved the TW-equation in cylindrical coordinates, but used different representations of the topographic profile. Mysak used the finite width sloping profile of *Figure 3.12a* (in which y is now the radial distance) Saint-Guily applied the parabolic depth profile, while Rhines, Buchwald & Melville and Hogg employed the power law

$$h(y) = \begin{cases} dy^\alpha, & a < y < a + r, \\ 1, & y > a + r, \end{cases}$$

with $\alpha > 0$; a is the radius of the island. To our knowledge, solutions of the TW-equation in the exterior of an elliptical island were not constructed so far.

The derivation of the relevant equations in elliptical coordinates ξ, η is given in section 2.5. We assume the isobaths to follow confocal ellipses, so that $h = h(\xi)$. The shore of the island will be given by $\xi = \xi_I$, the contour line of the outer edge of the shelf by $\xi = \xi_E$. The TW-equation is given by $(3.25)_1$. With the separation of variables solution

$$\psi(\xi, \eta) = F(\xi) \, e^{i(m\eta + \sigma t)}, \qquad m = 1, 2, 3, \ldots$$

(the + sign of σ is used because a right-bounded phase propagation is clockwise around the island) the boundary value problem becomes

$$(h^{-1} F')' + (\frac{m}{\sigma} (h^{-1})' - m^2 h^{-1}) F = 0, \qquad \zeta > \xi_I,$$

$$F = 0, \qquad\qquad\qquad\qquad\qquad\qquad \text{at} \quad \xi = \xi_I, \infty, \qquad (3.47)$$

$$[\![F]\!] = [\![F']\!] = 0, \qquad\qquad\qquad\qquad \text{at} \quad \xi = \xi_E.$$

With the exponential shelf profile

$$h(\xi) = \begin{cases} e^{b(\xi - \xi_E)}, & \xi_I \leq \xi \leq \xi_E, \\ 1, & \xi_E \leq \xi, \end{cases}$$

the solution of (3.47) satisfying the boundary conditions reads

$$F(\xi) = \begin{cases} e^{b(\xi - \xi_E)/2} \sin \lambda (\xi - \xi_I), & \xi_I \leq \xi \leq \xi_E, \\ A e^{-m\xi} & \xi_E \leq \xi, \end{cases}$$

with

$$\lambda^2 = \frac{mb}{\sigma} - m^2 - \frac{b^2}{4}.$$

The matching conditions at ξ_E determine the constant A and yield the eigenvalue equation

$$\frac{1}{\lambda} \tan \lambda (\xi_E - \xi_I) = - \frac{1}{m + \dfrac{b}{2}}, \qquad m = 1,2,3,\ldots. \qquad (3.48)$$

This is exactly analogous to the dispersion relation (3.44) for TW's of a straight shelf, except that the "wavenumber" m is quantized here due to the 2π-periodicity in η. (3.48) is an example of an infinite domain with a discrete spectrum. *Table 3.9* lists a selection of eigenfrequencies for various values of m, ξ_I, ξ_E and b.

r	ξ_I	A_E/A_I	b	Mode (m,n)			
				$(1,1)$	$(2,1)$	$(3,1)$	$(1,2)$
		2	3.30	0.232	0.358	0.405	0.006
0.99	2.65	3	2.09	0.317	0.408	0.399	0.088
		5	1.43	0.379	0.401	0.347	0.120
		2	2.48	0.285	0.396	0.410	0.076
0.5	0.549	3	1.69	0.354	0.410	0.374	0.105
		5	1.22	0.397	0.385	0.318	0.135
		2	1.88	0.336	0.411	0.388	0.096
0.1	0.100	3	1.38	0.384	0.398	0.341	0.123
		5	1.05	0.408	0.363	0.289	0.149

Table 3.9

Eigenfrequencies of the (m=1,n=1), (2,1), (3,1), (1,2) modes of TW's around an elliptic island according to (3.48). r is the aspect ratio (width to length) of the island, A_E and A_I are the semi-axes of the elliptic shelf boundary, $A_E = \cosh \xi_E$, and of the island $A_I = \cosh \xi_I$, and b is a topography parameter such that $h(\xi_I) = 0.1$.

4. The Method of Weighted Residuals

4.1 Application to the TW-equation

Construction of analytical solutions to the TW-equation (2.24) subject
to the no-flux boundary condition is possible only for some simple cases.
Even though the equation may still be separable when written in a special coordinate system, the solution of the emerging ordinary differential equation may either only be expressible in terms of special functions which are tedious to handle, or must be obtained numerically. With
realistic boundaries and non-vanishing curvature of the domain an exact
solution can hardly be found. In this chapter we therefore introduce a
procedure, by which equation (2.24) can be solved *approximately*. The method
consists of a reduction of the dimension of the mathematical problem by
a basis (shape) function expansion and is a variant of the *projection method*, the *spectral* or *modal method* and may also be considered a *generalized
separation of variables procedure*. Its advantage is that despite of its numerical intent, the method permits analytical techniques to be pursued farther than with classical numerical approaches.

There are several techniques by which the reduction of the dimensionality
of a boundary value problem can be achieved and then approximately solved. One is to derive the governing equations from a *Variational Principle*.
For the TW-equation this involves construction of a *functional* (Lagrang-
ian) in terms of the mass transport stream function; the TW-equation is
obtained as the Euler-Lagrange equation of this functional and the boundary condition would result from the natural boundary condition of the
variation of the functional. Ripa (1978) and Mysak (1985) proceed this
way. We use here (as we have done in Chapter 2 for the derivation of the
governing equations of a continuously stratified lake) the *Method of Weighted Residuals* (MWR). Both methods, in their essentials, are described in
Finlayson (1972). The MWR has already been applied to gravity waves by
Raggio & Hutter (1982), to topographic waves by Stocker & Hutter (1985,
1986 a,b) and to two-phase turbidity currents by Scheiwiller et al.
(1986).

The MWR and the variational principle in the function expansion approach
are related to the Method of Finite Elements (FE). One fundamental difference, however, consists in the fact that the domain of integration is
not partitioned into a number of elements in which linear or higher order
interpolation is performed. Rather than assuming the *local* functional
dependence within an individual element and then minimizing some global

measure, our model approach prescribes the *global* functional dependence along one dimension and maps the problem into the orthogonal subspace. This is achieved by a weighted integration of the equations along this dimension.

We consider the eigenvalue problem (2.24) formulated in the natural co-ordinate system shown in *Figure 4.1*,

$$\mathbb{D}\,\psi = 0, \quad \text{in} \quad \mathcal{D},$$
$$\mathbb{B}\,\psi = 0, \quad \text{in} \quad \partial\mathcal{D}, \tag{4.1}$$

with the definition of the differential and boundary operators $\mathbb{D}$ and $\mathbb{B}$, respectively

$$\mathbb{D} \equiv \frac{1}{J}\left[-i\sigma\left[\frac{\partial}{\partial s}(\frac{h^{-1}}{J}\frac{\partial}{\partial s}) + \frac{\partial}{\partial n}(h^{-1}\,J\,\frac{\partial}{\partial n})\right] + \frac{\partial h^{-1}}{\partial n}\,\frac{\partial}{\partial s} - \frac{\partial h^{-1}}{\partial s}\,\frac{\partial}{\partial n}\right],$$
$$\mathbb{B} \equiv 1. \tag{4.2}$$

$\sigma \equiv \omega/f$ is the non-dimensional frequency and $J = 1-Kn$, where K is the curvature of the thalweg.

Figure 4.1

Elongated lake and transverse section in a natural (s,n,z)-coordinate system. The thalweg axis $(n=0)$ may be a center of symmetry (not necessarily) and have curvature $K(s)$.

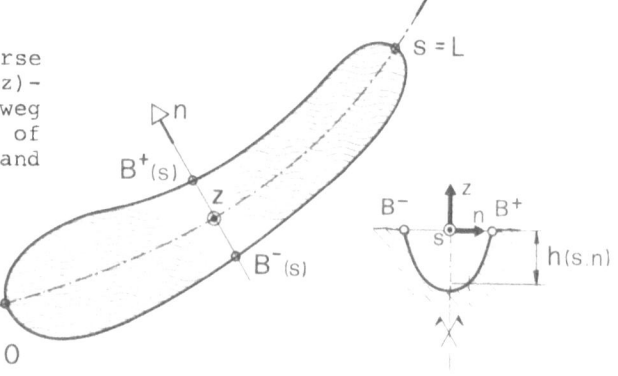

Let $\{P_\alpha(s,n)\}$ be a complete set of basis functions indexed by α, in terms of which the mass transport stream function $\psi(s,n)$ is expanded:

$$\psi(s,n) = \sum_{\alpha=1}^{N} P_\alpha(s,n)\,\psi_\alpha(s) \equiv P_\alpha \psi_\alpha. \tag{4.3}$$

Each basis function is weighted by a residue function $\psi_\alpha(s)$ which is as-sumed not to depend on the transverse coordinate n. All functional de-pendence on n is now incorporated in the preselected basis functions P_α, a general form of separation. Expansion (4.3) represents the exact solu-tion for a separable problem provided the basis functions are appropria-tely selected. For non-separable systems as (2.24) generally is, and for an arbitrary set $\{P_\alpha\}$ with $N < \infty$, the expansion is merely an approxima-tion. Clearly, fast convergence is anticipated so that truncation of (4.3) for very small N may furnish a sufficiently accurate solution.

The integration of (4.1) with an arbitrary bounded weighting function $\delta\phi(s,n)$ over the lake domain and along the shoreline, respectively, leads to the integral formulations

$$\int_D (\mathbb{D}\,\psi)\,\delta\phi\;da = 0, \qquad \oint_{\partial D}(\mathbb{B}\,\psi)\,\delta\phi\;d\ell = 0. \tag{4.4}$$

If (4.4) holds for any weighting function this is equivalent to (4.1) owing to the fundamental lemma of the Calculus of Variation (Courant & Hilbert, 1967). Expanding also the weighting function in terms of the complete set $\{Q_\beta\}$, viz.

$$\delta\phi(s,n) = \sum_{\beta=1}^{N} Q_\beta(s,n)\,\delta\phi_\beta(s) \equiv Q_\beta\,\delta\phi_\beta,$$

and inserting these expansions into (4.4) yields

$$\int_D (\mathbb{D}\,P_\alpha\,\psi_\alpha)\,Q_\beta\,\delta\phi_\beta\;da = 0, \quad \oint_{\partial D}(\mathbb{B}\,P_\alpha\,\psi_\alpha)\,Q_\beta\,\delta\phi_\beta\;d\ell = 0. \tag{4.5a,b}$$

The integration over the lake domain D can be split up into two integrations over either coordinates using $da = J\,dn\,ds$ for the area element in the natural coordinate frame. Further, the trivial form of the boundary operator $\mathbb{B} \equiv 1$ suggests the special choices

$$P_\alpha(s,B^\pm) = 0, \quad Q_\beta(s,B^\pm) = 0, \quad \text{for all } \alpha,\beta \tag{4.6}$$

such that the only contribution to (4.5b) arises from the ends of the lake.

Since the weighting functions are arbitrary, (4.5) can be replaced by

$$\left.\begin{array}{r} \displaystyle\int_{n=B^-}^{n=B^+}(\mathbb{D}\,P_\alpha\,\psi_\alpha)\,JQ_\beta\;dn = 0 \\[4pt] \psi_\alpha(s)\Big|_{s=0,L} = 0 \end{array}\right\} \quad \alpha,\beta = 1,\ldots,N. \tag{4.7}$$

The residue functions ψ_α depend only on s and are therefore extracted from the integration by carefully accounting for the effect of the differential operator $\mathbb{D}$ on $\psi_\alpha(s)$. Substituting (4.2) into (4.7), we obtain

$$0 = \int_{B^-}^{B^+}\Big[-i\sigma\,\{\underbrace{\frac{\partial}{\partial s}\Big[\frac{h^{-1}}{J}\,\frac{\partial}{\partial s}(P_\alpha\,\psi_\alpha)\Big]}_{(1)} + \underbrace{\frac{\partial}{\partial n}\Big[h^{-1}J\,\frac{\partial}{\partial n}(P_\alpha\,\psi_\alpha)\Big]}_{(2)}\}$$

$$+ \underbrace{\frac{\partial h^{-1}}{\partial n}\,\frac{\partial}{\partial s}(P_\alpha\,\psi_\alpha)}_{(3)} - \underbrace{\frac{\partial h^{-1}}{\partial s}\,\frac{\partial}{\partial n}(P_\alpha\,\psi_\alpha)}_{(4)}\Big]\,Q_\beta\;dn,$$

where use of the summation convention has been made. Each term in this expression will be evaluated separately. In the following deductions we need Leibnitz integration rule

$$\int\limits_{B^-}^{B^+} \frac{\partial F}{\partial s} G \, dn = \int\limits_{B^-}^{B^+} \frac{\partial}{\partial s}(FG) \, dn - \int\limits_{B^-}^{B^+} F \frac{\partial G}{\partial s} \, dn$$

$$= \frac{\partial}{\partial s} \int\limits_{B^-}^{B^+} FG \, dn - (FG)\Big|_{B^+} \cdot \frac{\partial B^+}{\partial s} + (FG)\Big|_{B^-} \cdot \frac{\partial B^-}{\partial s} - \int\limits_{B^-}^{B^+} F \frac{\partial G}{\partial s} \, dn,$$

where F and G are arbitrary, differentiable functions of s and n.

With these preliminaries the terms (1) to (4) can be evaluated. The rule of transformation is to remove differentiations of the topography h as far as possible, which can be achieved by integration by parts:

Term (1):

$$(1) = \int \frac{\partial}{\partial s}\Big(\frac{h^{-1}}{J} \frac{\partial}{\partial s}(P_\alpha \psi_\alpha)\Big) Q_\beta \, dn$$

$$= \frac{\partial}{\partial s} \int \frac{h^{-1}}{J} \frac{\partial}{\partial s}(P_\alpha \psi_\alpha) Q_\beta \, dn - \int \frac{h^{-1}}{J} \frac{\partial}{\partial s}(P_\alpha \psi_\alpha) \frac{\partial Q_\beta}{\partial s} \, dn$$

$$= \frac{\partial}{\partial s}\Big[\psi_\alpha \int \frac{h^{-1}}{J} \frac{\partial P_\alpha}{\partial s} Q_\beta \, dn + \frac{\partial \psi_\alpha}{\partial s} \int \frac{h^{-1}}{J} P_\alpha Q_\beta \, dn \Big]$$

$$- \psi_\alpha \int \frac{h^{-1}}{J} \frac{\partial P_\alpha}{\partial s} \frac{\partial Q_\beta}{\partial s} \, dn - \frac{\partial \psi_\alpha}{\partial s} \int \frac{h^{-1}}{J} P_\alpha \frac{\partial Q_\beta}{\partial s} \, dn$$

$$= \psi_\alpha\Big[\frac{\partial}{\partial s} \int \frac{h^{-1}}{J} \frac{\partial P_\alpha}{\partial s} Q_\beta \, dn - \int \frac{h^{-1}}{J} \frac{\partial P_\alpha}{\partial s} \frac{\partial Q_\beta}{\partial s} \, dn \Big]$$

$$+ \frac{\partial \psi_\alpha}{\partial s}\Big[\int \frac{h^{-1}}{J} \frac{\partial P_\alpha}{\partial s} Q_\beta \, dn + \frac{\partial}{\partial s} \int \frac{h^{-1}}{J} P_\alpha Q_\beta \, dn$$

$$- \int \frac{h^{-1}}{J} P_\alpha \frac{\partial Q_\beta}{\partial s} \, dn \Big] + \frac{\partial^2 \psi_\alpha}{\partial s^2} \int \frac{h^{-1}}{J} P_\alpha Q_\beta \, dn,$$

Term (2):

$$(2) = \int \frac{\partial}{\partial n}\Big[h^{-1} J \frac{\partial}{\partial n}(P_\alpha \psi_\alpha) \Big] Q_\beta \, dn$$

$$= - \int h^{-1} J \frac{\partial}{\partial n}(P_\alpha \psi_\alpha) \frac{\partial Q_\beta}{\partial n} \, dn$$

$$= - \psi_\alpha \int h^{-1} J \frac{\partial P_\alpha}{\partial n} \frac{\partial Q_\beta}{\partial n} \, dn,$$

Term (3):

$$(3) = \int \frac{\partial h^{-1}}{\partial n} \frac{\partial}{\partial s}(P_\alpha \psi_\alpha) Q_\beta \, dn$$

$$= - \int \frac{\partial}{\partial n}\Big(\frac{\partial}{\partial s}(P_\alpha \psi_\alpha) Q_\beta\Big) \cdot h^{-1} \, dn$$

$$= - \psi_\alpha \int h^{-1} \frac{\partial}{\partial n}\Big(\frac{\partial P_\alpha}{\partial s} Q_\beta\Big) \, dn - \frac{\partial \psi_\alpha}{\partial s} \int h^{-1} \frac{\partial}{\partial n}(P_\alpha Q_\beta) \, dn,$$

Term (4):

$$(4) = -\int \frac{\partial h^{-1}}{\partial s} \frac{\partial}{\partial n} (P_\alpha \psi_\alpha) Q_\beta \, dn$$

$$= - \frac{\partial}{\partial s} \int h^{-1} \frac{\partial}{\partial n}(P_\alpha \psi_\alpha) Q_\beta \, dn + \int h^{-1} \frac{\partial}{\partial s} \left(\frac{\partial}{\partial n}(P_\alpha \psi_\alpha) Q_\beta\right) dn$$

$$= - \frac{\partial \psi_\alpha}{\partial s} \int h^{-1} \frac{\partial P_\alpha}{\partial n} Q_\beta \, dn - \psi_\alpha \frac{\partial}{\partial s} \int h^{-1} \frac{\partial P_\alpha}{\partial n} Q_\beta \, dn$$

$$+ \psi_\alpha \int h^{-1} \frac{\partial}{\partial s}(\frac{\partial P_\alpha}{\partial n} Q_\beta) \, dn + \frac{\partial \psi_\alpha}{\partial s} \int h^{-1} \frac{\partial P_\alpha}{\partial n} Q_\beta \, dn.$$

Parenthetically we may remark that the process of this evaluation is
more complex when the basis functions are not restricted by the condition that they vanish along the shore, because further integration by
parts is necessary in that case. (4.7) thus takes the form

$$0 = - i\sigma \left[\frac{\partial^2 \psi_\alpha}{\partial s^2} \left[\int \frac{h^{-1}}{J} P_\alpha Q_\beta \, dn \right] \right.$$

$$+ \frac{\partial \psi_\alpha}{\partial s} \left[\int \frac{h^{-1}}{J} \frac{\partial P_\alpha}{\partial s} Q_\beta \, dn + \frac{\partial}{\partial s} \int \frac{h^{-1}}{J} P_\alpha Q_\beta \, dn - \int \frac{h^{-1}}{J} P_\alpha \frac{\partial Q_\beta}{\partial s} \, dn \right]$$

$$+ \psi_\alpha \left[\frac{\partial}{\partial s} \int \frac{h^{-1}}{J} \frac{\partial P_\alpha}{\partial s} Q_\beta \, dn - \int \frac{h^{-1}}{J} \frac{\partial P_\alpha}{\partial s} \frac{\partial Q_\beta}{\partial s} \, dn - \int h^{-1} J \frac{\partial P_\alpha}{\partial n} \frac{\partial Q_\beta}{\partial n} \, dn \right] \qquad (4.8)$$

$$+ \frac{\partial \psi_\alpha}{\partial s} \left[- \int h^{-1} \frac{\partial P_\alpha}{\partial n} Q_\beta \, dn - \int h^{-1} P_\alpha \frac{\partial Q_\beta}{\partial n} \, dn \right]$$

$$+ \psi_\alpha \left[\int h^{-1} \frac{\partial P_\alpha}{\partial n} \frac{\partial Q_\beta}{\partial s} \, dn - \int h^{-1} \frac{\partial P_\alpha}{\partial s} \frac{\partial Q_\beta}{\partial n} \, dn - \frac{\partial}{\partial s} \int h^{-1} \frac{\partial P_\alpha}{\partial n} Q_\beta \, dn \right],$$

and the integrals are understood as $\int_{B^-}^{B^+} ..$ (4.8) can be written in the
form

$$\left. \begin{array}{r} \mathbf{M}_{\beta\alpha} \psi_{\alpha\cdot} = 0 \\ \psi_\alpha = 0 \end{array} \right\} \alpha, \beta = 1, \ldots, N \quad \left\{ \begin{array}{l} 0 < s < L, \\ s = 0, L, \end{array} \right. \qquad (4.9)$$

with the matrix operator elements

$$\mathbf{M}_{\beta\alpha} = - i\sigma \left[M_{\beta\alpha}^{00} \frac{d^2}{ds^2} + \left(\frac{dM_{\beta\alpha}^{00}}{ds} + M_{\beta\alpha}^{10} - M_{\beta\alpha}^{01}\right) \frac{d}{ds} + \left(\frac{dM_{\beta\alpha}^{10}}{ds} - M_{\beta\alpha}^{11} - M_{\beta\alpha}^{22}\right) \right]$$

$$- (M_{\beta\alpha}^{20} + M_{\beta\alpha}^{02}) \frac{d}{ds} - \left(\frac{dM_{\beta\alpha}^{20}}{ds} + M_{\beta\alpha}^{12} - M_{\beta\alpha}^{21}\right) \qquad (\alpha, \beta = 1, \ldots, N). \qquad (4.10)$$

The matrix elements $M_{\beta\alpha}^{ij}$ represent quadrature formulae in the transverse
direction, explicitly:

$$M_{\beta\alpha}^{00} = \int h^{-1} J^{-1} P_\alpha Q_\beta \, dn,$$

$$M_{\beta\alpha}^{10} = \int h^{-1} J^{-1} \frac{\partial P_\alpha}{\partial s} Q_\beta \, dn, \qquad M_{\beta\alpha}^{01} = \int h^{-1} J^{-1} P_\alpha \frac{\partial Q_\beta}{\partial s} \, dn,$$

$$M_{\beta\alpha}^{20} = \int h^{-1} \frac{\partial P_\alpha}{\partial n} Q_\beta \, dn, \qquad M_{\beta\alpha}^{02} = \int h^{-1} P_\alpha \frac{\partial Q_\beta}{\partial n} \, dn, \qquad (4.11)$$

$$M_{\beta\alpha}^{11} = \int h^{-1} J^{-1} \frac{\partial P_\alpha}{\partial s} \frac{\partial Q_\beta}{\partial s} \, dn, \qquad M_{\beta\alpha}^{22} = \int h^{-1} J \frac{\partial P_\alpha}{\partial n} \frac{\partial Q_\beta}{\partial n} \, dn,$$

$$M_{\beta\alpha}^{12} = \int h^{-1} \frac{\partial P_\alpha}{\partial s} \frac{\partial Q_\beta}{\partial n} \, dn, \qquad M_{\beta\alpha}^{21} = \int h^{-1} \frac{\partial P_\alpha}{\partial n} \frac{\partial Q_\beta}{\partial s} \, dn.$$

The individual components $M_{\beta\alpha}^{ij}$ in (4.11) are known functions of s and depend on the topography of the lake, h, on the metric of the natural coordinate system, $J(s,n)$, on the shape of the lake shore, $B^{\pm}(s)$, and on the sets of basis functions $\{P_\alpha(s,n)\}$ and $\{Q_\beta(s,n)\}$.

Notice that (4.9) is only meaningful as long as all·entries of the matrices (4.11) are bounded. Since J and J^{-1} are both regular, this means that the basis functions P_α and Q_β must be chosen such that the combinations $h^{-1} P_\alpha Q_\beta$, $h^{-1} \partial P_\alpha/\partial s \, Q_\beta$, etc. arising in (4.11) are integrable. For $h > 0$ no difficulties arise, however, when $h = 0$ along the shore the functions P_α, Q_β must be taken from a set of which the near-shore behavior is dictated by that of h. This is a drawback of this method and restricts it essentially to profiles with finite shore depth.

Equations (4.9) form a system of coupled one-dimensional differential equations that replace the single two-dimensional boundary-value problem (4.1). These two formulations are presumed to be equivalent provided (i) the sets of basis functions are complete in $[B^-, B^+]$ and (ii) $N = \infty$. The selected order N of the system sets a natural bound to the variability of the approximate solution as well as to its quality. At a first glance the MWR seems to leave us with a more complicated task. Finite-difference calculations, however, have indicated numerical difficulties such as slow convergence, particularly for complicated topographies and for large wavenumbers (Bäuerle, 1986). This semi-analytical procedure may thus well prove advantageous in achieving a better physical understanding.

4.2 Symmetrization

More insight into the structure of the operator (4.10) is gained when the physical configuration exhibits symmetry with respect to the axis $n = 0$. Such a symmetry may exist for channels and it often applies approximately for elongated, narrow lakes. The symmetrization is also motivated by the fact that solutions found for circular and elliptic basins, indicate that the phase rotates counterclockwise and the stream function continuously changes its symmetry with respect to the symmetry axis of the lake[*]. As a consequence a split into symmetric and skew-symmetric basis functions is appropriate. We shall, for the purpose of studying channels and basins which have a symmetry axis, formulate problem (4.9) in a symmetrized version. To this end, the functions P_α, Q_β, J and J^{-1} are symmetrized by introducing the decompositions

$$f(s,n) = f^+(s,n) + f^-(s,n),$$

$$f^+(s,n) = f^+(s,-n), \qquad (4.12)$$

$$f^-(s,n) = -f^-(s,-n).$$

This decomposition is applied to the matrix elements $M_{\beta\alpha}^{ij}$ in (4.11); the important result here is

$$M_{\beta\alpha}^{00} = M_{\beta\alpha}^{00++} + M_{\beta\alpha}^{00--} + M_{\beta\alpha}^{00-+} + M_{\beta\alpha}^{00+-} \quad (\alpha,\beta = 1,\ldots,N)$$

$$= \int h^{-1}(J^{-1})^+ P_\alpha^+ Q_\beta^+ \, dn + \int h^{-1}(J^{-1})^+ P_\alpha^- Q_\beta^- \, dn$$

$$+ \int h^{-1}(J^{-1})^- P_\alpha^- Q_\beta^+ \, dn + \int h^{-1}(J^{-1})^- P_\alpha^+ Q_\beta^- \, dn , \qquad (4.13)$$

$$M_{\beta\alpha}^{20} = M_{\beta\alpha}^{20-+} + M_{\beta\alpha}^{20+-}$$

$$= \int h^{-1} \frac{\partial P_\alpha^-}{\partial n} Q_\beta^+ \, dn + \int h^{-1} \frac{\partial P_\alpha^+}{\partial n} Q_\beta^- \, dn ,$$

with analogous expressions for $M_{\beta\alpha}^{22}$ and $M_{\beta\alpha}^{02}$ respectively. It has been assumed above that $h^- = 0$ (symmetric depth profile), and the integration is from $B^- = -\frac{1}{2}B(s)$ to $B^+ = \frac{1}{2}B(s)$. Because the basis funtions P_α and Q_β are decomposed according to (4.12) the expansion (4.3) of the solution $\psi(s,n)$ must be replaced by

---

[*] *If in Figure 3.3 the long axis is identified with the s-axis it is seen that for $t = 0$ the mass transport stream function ψ is skew-symmetric and for $t = T/4$ it is symmetric.*

$$\psi(s,n) = P_\alpha^+(s,n)\,\psi_\alpha^+(s) + P_\alpha^-(s,n)\,\psi_\alpha^-(s),$$

where the $\pm$ superscripts on ψ_α indicate merely affiliation to the individual $P_\alpha^\pm$. In vector notation the stream function reads

$$\underset{\sim}{\psi} = (\psi_1^+,\ldots,\psi_N^+;\ \psi_1^-,\ldots,\psi_N^-) = (\underset{\sim}{\psi}^+;\underset{\sim}{\psi}^-),$$

and the matrices (4.13) take the form

$$\underset{\sim}{M}^{00} = \begin{bmatrix} \underset{\sim}{M}^{00++} & \underset{\sim}{M}^{00-+} \\ \underset{\sim}{M}^{00+-} & \underset{\sim}{M}^{00--} \end{bmatrix},\qquad \underset{\sim}{M}^{20} = \begin{bmatrix} \underset{\sim}{0} & \underset{\sim}{M}^{20-+} \\ \underset{\sim}{M}^{20+-} & \underset{\sim}{0} \end{bmatrix},\ \text{etc.}$$

With this notation the differential equations (4.9) read

$$\left(-i\sigma \begin{bmatrix} \underset{\sim}{M}^{++} & \underset{\sim}{M}^{-+} \\ \underset{\sim}{M}^{+-} & \underset{\sim}{M}^{--} \end{bmatrix} + \begin{bmatrix} \underset{\sim}{0} & \underset{\sim}{N}^{-+} \\ \underset{\sim}{N}^{+-} & \underset{\sim}{0} \end{bmatrix}\right)\begin{pmatrix} \underset{\sim}{\psi}^+ \\ \underset{\sim}{\psi}^- \end{pmatrix} = 0, \qquad (4.14)$$

with the matrix operators $\underset{\sim}{M}$ and $\underset{\sim}{N}$ the particular form of which is unimportant in the ensuing arguments.

The coupling of the solution vectors ψ^+ and ψ^- is induced by the off-diagonal operators $\underset{\sim}{M}^{-+}$, $\underset{\sim}{M}^{+-}$ and $\underset{\sim}{N}^{-+}$, $\underset{\sim}{N}^{+-}$ respectively. The former are due to curvature and vanish when $K = 0$. The latter originate from the vector product in equation (2.22) and express the effect of the Coriolis force. The restriction to only symmetric basis functions reduces (4.14) to two decoupled equations. This obviously corresponds to the claim that both terms of the sum of equation (2.22) be individually zero. On imposing the boundary condition this implies $\psi \equiv 0$, c.f. section 2.2. It suggests that the approximate system requires a set of basis functions containing both symmetric and antisymmetric functions if qualitatively correct results are to emerge.

The remainder of this monograph will almost exclusively be concerned with the solution of equation (4.14) in various different domains.

5. Topographic waves in infinite channels

In this chapter properties of topographic waves in infinite channels will be discussed. The first four sections deal with our own solution procedure using the model equations (4.9) and follow Stocker & Hutter (1986a) with additions from Stocker & Hutter (1985). In section 5.5 solutions obtained with analytical and numerical finite difference techniques (Gratton, 1983, Gratton & LeBlond, 1986, Bäuerle, 1986) will be investigated. Finally section 5.6 presents solutions in curved channels.

5.1 Basic concept

The suitability of the approximate model equations (4.9) deduced with the MWR is now tested using a straight, infinite and symmetric channel with a topography of the form

$$h(s,n) = h_0(s)(1+\varepsilon - \left|\frac{2n}{B(s)}\right|^q), \tag{5.1}$$

where ε is a sidewall and q a topography parameter, see *Figure 5.1*, which provides the possibility of modelling both concave ($q > 1$) and convex ($q < 1$) transverse depth profiles. The sidewall parameter ε has been in-

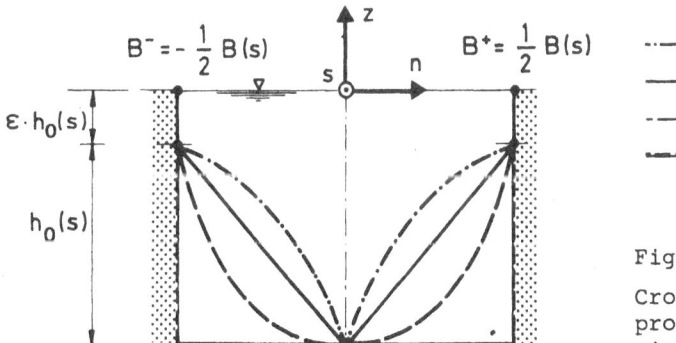

Figure 5.1

Cross-sectional depth profiles of a symmetric channel.

troduced in order that all matrix elements (4.11) take finite values. The complete sets of basis functions $\{P_\alpha\}$ and $\{Q_\beta\}$ will be chosen to be identical (Galerkin procedure) with the symmetric and skew-symmetric parts reading, see *Figure 5.2*,

$$P_\alpha^+(s,n) = \cos\left(\pi(\alpha-\frac{1}{2})\frac{2n}{B(s)}\right),$$

$$(\alpha = 1,\ldots,N). \tag{5.2}$$

$$P_\alpha^-(s,n) = \sin\left(\pi\alpha\frac{2n}{B(s)}\right),$$

Here, P_α^+ and P_α^- arise in pairs; N thus characterizes a model consisting of 2N basis functions. These satisfy the boundary conditions (4.6) along the shoreline $n = \pm \frac{1}{2} B(s)$. Substituting (5.1) and (5.2) into (4.11) and assuming B(s) to be constant*) it is seen that

$$M_{\beta\alpha}^{00} = B \, h_0^{-1} \, K_{\beta\alpha}^{00}, \qquad M_{\beta\alpha}^{22} = B^{-1} \, h_0^{-1} \, K_{\beta\alpha}^{22},$$

$$M_{\beta\alpha}^{20} = h_0^{-1} \, K_{\beta\alpha}^{20}, \qquad M_{\beta\alpha}^{02} = h_0^{-1} \, K_{\beta\alpha}^{02},$$

while the elements with the superscripts 10, 01, 11, 12 and 21 all vanish.

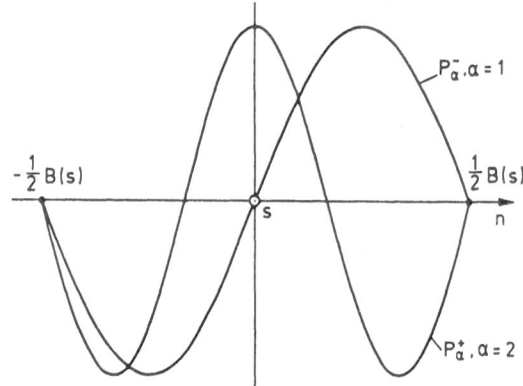

Figure 5.2

Symmetric and skew-symmetric basis functions P_α.

The dimension-free matrix elements $K_{\beta\alpha}^{ij}$ depend on ε and q and straightforward calculation leads to the expressions for these:

$$K_{\beta\alpha}^{00++} = \int h_*^{-1} \cos \pi (\alpha - \tfrac{1}{2}) y \cos \pi(\beta - \tfrac{1}{2}) y \, dy,$$

$$K_{\beta\alpha}^{00--} = \int h_*^{-1} \sin \pi\alpha y \, \sin \pi\beta y \, dy,$$

$$K_{\beta\alpha}^{22++} = 4 \pi^2 (\alpha - \tfrac{1}{2}) (\beta - \tfrac{1}{2}) \int h_*^{-1} \sin \alpha (\alpha - \tfrac{1}{2}) y \, \sin \pi (\beta - \tfrac{1}{2}) y \, dy,$$

$$K_{\beta\alpha}^{22--} = 4 \pi^2 \alpha\beta \int h_*^{-1} \cos \pi\alpha y \, \cos \pi\beta y \, dy,$$

$$K_{\beta\alpha}^{20+-} = - 2\pi (\alpha - \tfrac{1}{2}) \int h_*^{-1} \sin \pi (\alpha - \tfrac{1}{2}) y \, \sin \pi \, \beta y \, dy,$$

$$K_{\beta\alpha}^{20-+} = 2\pi\alpha \int h_*^{-1} \cos \pi\alpha y \, \cos \pi (\beta - \tfrac{1}{2}) y \, dy,$$

$$K_{\beta\alpha}^{02+-} = 2\pi\alpha \int h_*^{-1} \cos \pi (\alpha - \tfrac{1}{2}) y \, \cos \pi \, \beta y \, dy,$$

$$K_{\beta\alpha}^{02-+} = - 2\pi (\beta - \tfrac{1}{2}) \int h_*^{-1} \sin \pi\alpha y \, \sin \pi (\beta - \tfrac{1}{2}) y \, dy,$$

(5.3)

with $h_* = 1 + \varepsilon - y^q$ and the integration is meant to be from $y = 0$ to $y = 1$. The numerical evaluation of these elements was performed on a CYBER com-

---

*) This assumption is not necessary and the operator $\underset{\sim}{I\!M}$ for $\partial B/\partial s \neq 0$ is given in Stocker & Hutter (1985).

puter using IMSL-library subroutines. The results for q = 2 and ε = 0.05 are listed in Appendix B and a more complete list can be found in Stokker & Hutter (1985). (4.10) takes the form

$$\mathbb{K} \equiv Bh_0 \, \mathbb{M} = -i\sigma\left[B^2 \, \underset{\sim}{K}^{00} \frac{d^2}{ds^2} - B^2(h_0^{-1}\frac{dh_0}{ds}) \, \underset{\sim}{K}^{00} \frac{d}{ds} - \underset{\sim}{K}^{22}\right]$$
$$- B(\underset{\sim}{K}^{20} + \underset{\sim}{K}^{02}) \frac{d}{ds} + B(h_0^{-1}\frac{dh_0}{ds}) \, \underset{\sim}{K}^{20}. \qquad (5.4)$$

This operator has constant coefficients whenever the depth-profile is constant or exponential with respect to the basin axis. For an infinite channel, however, we prefer $h_0(s)$ = constant. A carrier-wave ansatz

$$\underset{\sim}{\psi} = (\underset{\sim}{\psi}^+; \underset{\sim}{\psi}^-) = e^{iks/L}(c_1,...,c_N;c_{N+1},...,c_{2N}) = e^{iks/L}\underset{\sim}{c}, \qquad (5.5)$$

with a dimensionless complex-valued wavenumber k, Im(k) ≠ 0 is meaningful in semi-infinite and finite channels, and a length L is then appropriate. With (5.4) and (5.5) the symmetrized form of (4.9) reduces to a system of algebraic equations

$$\underset{\sim}{C}\underset{\sim}{c} = 0,$$
$$\underset{\sim}{C} = \begin{bmatrix} \sigma((rk)^2 \, \underset{\sim}{K}^{00++} + \underset{\sim}{K}^{22++}) & -(rk)(\underset{\sim}{K}^{20-+} + \underset{\sim}{K}^{02-+}) \\ -(rk)(\underset{\sim}{K}^{20+-} + \underset{\sim}{K}^{02+-}) & \sigma((rk)^2\underset{\sim}{K}^{00--} + \underset{\sim}{K}^{22--}) \end{bmatrix}, \qquad (5.6)$$

in which the aspect ratio parameter r = B/L has been introduced. Notice that r and k enter only through the product rk, suggesting that solutions for r = 1 only need to be constructed. $\underset{\sim}{C}$ is a (2N × 2N)-matrix and depends on σ and k. Equation (5.6) admits a non-trivial solution vector $\underset{\sim}{c}$ if and only if

$$\det \underset{\sim}{C}(\sigma,k) - 0. \qquad (5.7)$$

This characteristic equation forms the *dispersion relation* σ(k) for topographic Rossby waves in a straight infinite channel. It is a polynomial equation of order 2N in $(rk)^2$ with real coefficients. For each frequency a Nth order model, therefore, yields 4N wavenumbers counting complex conjugates and pairs having opposite signs.

Let k_γ (γ = 1,..., 4N) be a root of (5.7) corresponding to a frequency σ and let c_γ, $(c_{\alpha\gamma})$ be the associated eigenvector (component) of (5.6). A general channel solution ψ(s,n,t) can then be written as

$$\psi(s,n,t) = e^{-i\sigma ft} \sum_{\gamma=1}^{4N} e^{ik_\gamma s/L} d_\gamma\left[\sum_{\alpha=1}^{N} P_\alpha^+(s,n) \, c_{\alpha\gamma} + \sum_{\alpha=N+1}^{2N} P_{\alpha-N}^-(s,n) \, c_{\alpha\gamma}\right], \qquad (5.8)$$

in which solutions belonging to individual k occur in a linear combination by an arbitrary complex vector $\underline{d}$, (d_γ). Representation (5.8) is an appropriate solution in a straight infinite channel. For this particular configuration problem (2.22) is separable, the coefficients of the separated differential equation, however, are non-constant, and for very special topographies exact solutions can be obtained, see later sections 5.2 and 5.5. The MWR probably offers more freedom in modelling the channel topography, because improved accuracy can be obtained by higher-order models, and convergence is expected.

5.2 Dispersion relation

Solutions of (5.7) may be plotted schematically for a first-order model, $N = 1$, in a $(Re(k), Im(k), \sigma)$-coordinate system, see *Figure 5.3*. This is a model which uses one symmetric and one skew-symmetric basis function of the form (5.2) and is of lowest possible order. Its graph is symmetric with respect to both axes $Re(k) = 0$ and $Im(k) = 0$. Three regimes 1, 2, 3 can be distinguished where the wavenumbers k take real, complex and purely imaginary values, respectively. *Table 5.1* gives the periods at which the individual regimes join for different topography and sidewall parameters. In regime 1 all wavenumbers k are real and, therefore, represent physically possible channel solutions. Evidently in regime 1, there exists for each frequency a long and a short wave. This is typical of Rossby waves and has also been observed for *shelf waves* in chapter 3, provided the slope parameter $S = h'/h$ was bounded in the domain. This is so also for channels:

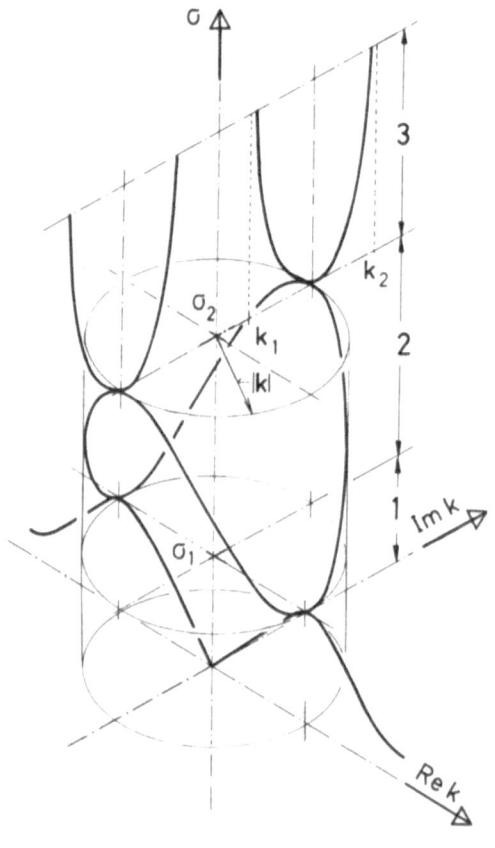

Figure 5.3

Schematic plot of the complex dispersion relation $\sigma(k)$ for an infinite channel with $\varepsilon = 0.05$ and $q = 0.5$ in a first-order model. In regime 1, k is real; in regime 2, it is complex with the constant modulus $|k|$; and in regime 3, k is purely imaginary, taking asymptotic values k_1 and k_2 for large σ.

in other words it can be proven that existence of a wavenumber $|k| = |k_0| < \infty$ such that $c_g = d\sigma/dk = 0$ is guaranteed only if h'/h is bounded everywhere across the channel width (see Appendix C). At this critical wavenumber no energy is transported along the channel. This corresponds roughly to wavelengths of about 0.5 ... 1 of the channel width, and the periods are listed in *Table 5.1*. It is also worth noting that Re(k) can have both signs. This is in contrast to planetary Rossby waves which are due to the β-effect (Holton, 1979) or Rossby waves on the continental shelf (Le Blond & Mysak, 1980), the reason being that here h'/h changes sign in the channel. So, such configurations enable topographic Rossby waves to propagate in *both* directions. In either case, as an effect of the Coriolis force, the structure of the wave on the Northern hemisphere is right-bounded with respect to the direction of phase propagation. The dispersion relation (5.7) contains only even powers of σ such that (5.7) is independent of the sign of σ. It is a convention that the sign of f (positive on the Northern hemisphere) determines the sign of the non-dimensional frequency σ.

q	$T_1[h]$		$T_2[h]$		$|k|$	
	ε = 0.05	ε = 0.10	ε = 0.05	ε = 0.10	ε = 0.05	ε = 0.10
0.5	52.8	58.3	10.5	11.8	6.6	5.9
1.0	60.5	64.3	13.2	14.4	6.9	6.2
2.0	83.0	88.2	22.0	22.6	6.8	6.3
5.0	174	199	58.2	61.8	6.1	5.8

Table 5.1
Periods and corresponding wavenumbers in a first-order model, which separate the regimes, depending on topography and sidewall parameters q and ε, respectively. The period T is calculated using T = 16.9 h/σ corresponding to 45° latitude. At T_1 no wave energy is transported.

The structure of the stream function depends upon the frequency range. Small frequencies (regime 1) favour periodic patterns along the channel. Waves with intermediate frequencies of order 1 (regime 2) have a mixed periodic-exponential structure and do not represent possible solutions in an infinite channel. At frequencies σ > 1 (regime 3) the solutions grow or decay exponentially. For later use, the union of the three regimes of the dispersion relation in *Figure 5.3* will be called a *mode unit*.

Let us proceed to the second-order model; it furnishes 8 wavenumbers to each frequency and its dispersion relation consists of two interlocking mode units, see *Figure 5.4*. Thus there are now two branches with real, complex and imaginary k, respectively. The relative size of the mode units and their spatial positions within the (k,σ)-coordinate system depend crucially upon the topography. The cylindrical surface of the first-order model degenerates to the smaller bell-shaped surface, i.e. |k| now depends on the frequency. The second mode unit forms an outer shell, which here has the form of a cone. Physically possible solutions for the infinite channel exist in regime 1 for both mode units and in regime 2 only for the first mode unit. The qualitative shape of the dispersion relation for an Nth-order model can be guessed from *Figures 5.3 and 5.4*. The modulus |k| is plotted for a third-order model in *Figure 5.5*, demonstrating clearly the addition of the next mode unit.

Summarizing the main points, we state the following remarks:

- The dispersion relation of an Nth order model consists of N mode units each of which has 3 regimes, in which wavenumbers are real, complex or imaginary.

- Solutions for infinite channels, which are physically meaningful, can only be constructed for wavenumbers k which are real. Therefore, when h'/h is bounded, there exist maximum frequencies, for which channel solutions may occur (see *Table 5.1*). At these maxima energy cannot propagate; for smaller k's group and phase velocities are unidirectional, for larger k's they are antidirectional.

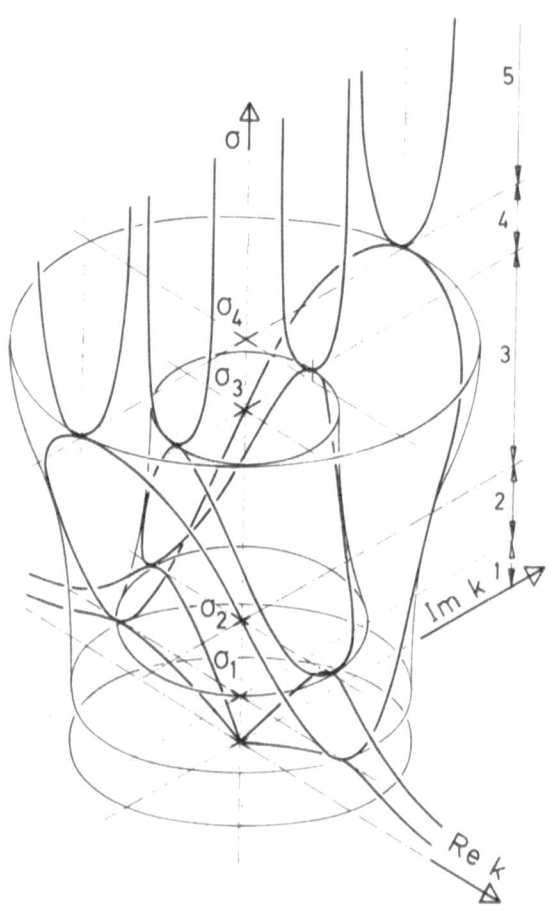

Figure 5.4

Schematic plot of the complex dispersion relation σ(k) for an infinite channel with ε = 0.05 and q = 0.5 in a second-order model. Five regimes with respect to σ can be differentiated.

- In domains, which are of finite extent also in the s-direction (lakes), solutions can be constructed with real, complex or imaginary wavenumbers k. Their spatial dependence is either periodic, periodic exponential or exponential.

- From this point of view, lake solutions occur for all $\sigma \in (0, \infty)$. However it must be remembered, that in section 2.3 a low-frequency approximation $\omega^2 \ll f^2$ was made. Therefore, physical applications of results with $|\sigma| > 1$ may be dubious.

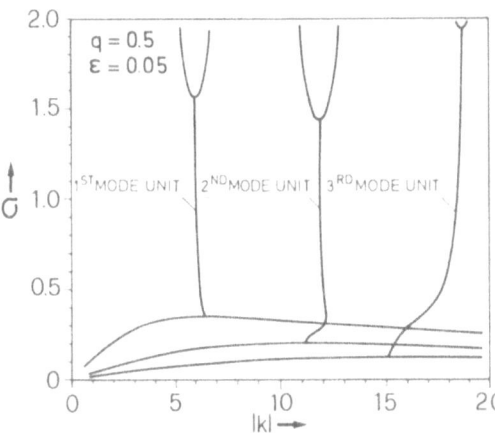

Figure 5.5

Modulus k of the third-order-dispersion relation for an infinite channel, $q = 0.5$, $\varepsilon = 0.05$.

The reader should be alerted to the fact, that these properties are tied to the existence of a finite $k = k_0$, where the purely real dispersion relation branches off to become complex.

The MWR is an approximate approach, and therefore convergence properties are expected. These are studied for the real branches of the dispersion relation. Figure 5.6 summarizes the results. The dispersion relation for $N = 3$ differs only slightly from that of the second-order model. The corrections of the second mode unit when increasing the order are also shown; however, for a statement on convergence a 4th-order model would be needed. Convergence is not uniform in k, being better for small k than for large k; furthermore, it is better for convex ($q = 0.5$, Figure 5.6a) than for concave ($q = 2$, Figures 5.6b,c) topographies, which is unfortunate as the latter are more realistic. Calculations have shown that the side-wall parameter ε does not influence convergence appreciably (Figure 5.6c).

The quality of the MWR-approximation is more obvious when the dispersion relation is compared with that of an exact solution as in Figure 5.7. The

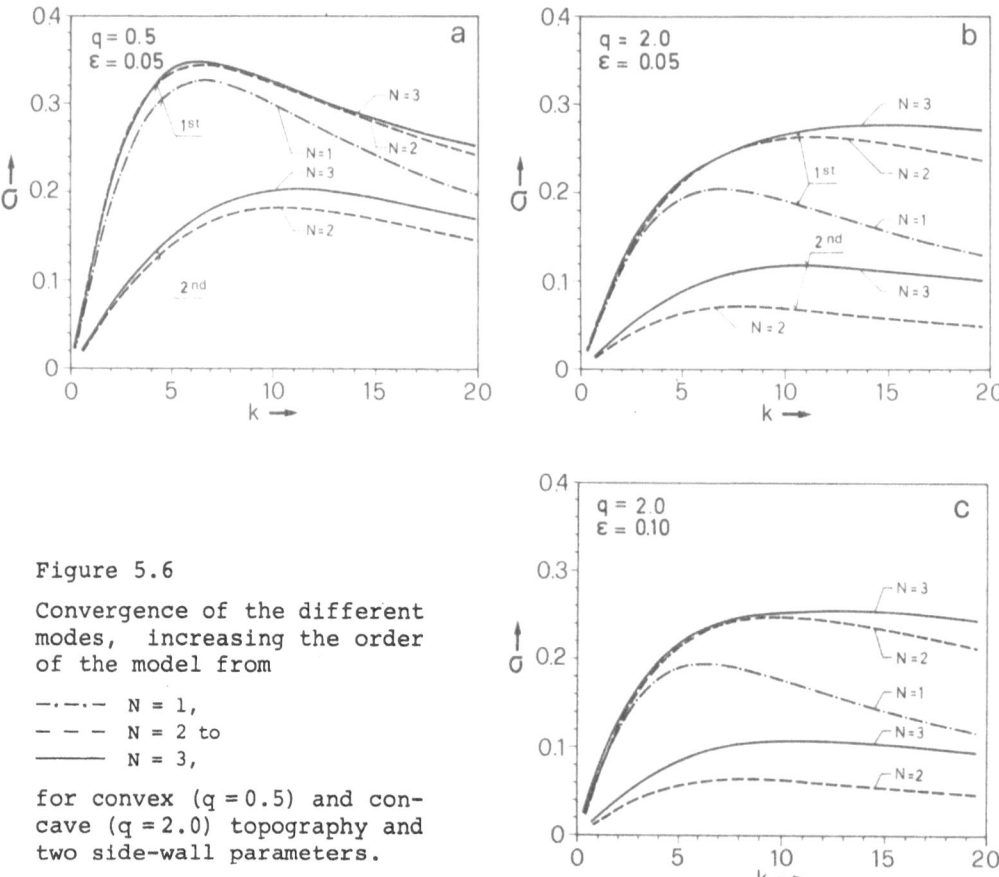

Figure 5.6

Convergence of the different modes, increasing the order of the model from

—·—·— N = 1,
— — — N = 2 to
———— N = 3,

for convex (q = 0.5) and concave (q = 2.0) topography and two side-wall parameters.

simple configuration of a straight channel leads to separable equations; these are easy to integrate provided the depth profile is piecewise exponential as indicated in the inset of *Figure 5.7*. The dispersion relation $\sigma(k)$ evolves from the matching conditions of the stream function within the channel. As *Figure 5.7a* demonstrates, the approximate dispersion curves calculated by the MWR applied to the same depth profile converge fast to the exact dispersion relation for the first mode. N = 2 already represents a satisfactory approximation within a few percent Convergence of the second mode is slower, as stated earlier. For steeper depth profiles, *Figure 5.7b*, convergence is significantly slower and higher-order models may be required. But it also appears that the selected set of basis functions is not best for such configurations, as wave activity is concentrated at the shore.

Figure 5.8 shows the influence of the variation of the topography parameter q in a first- and third-order model. Comparison of *Figures 5.8a*

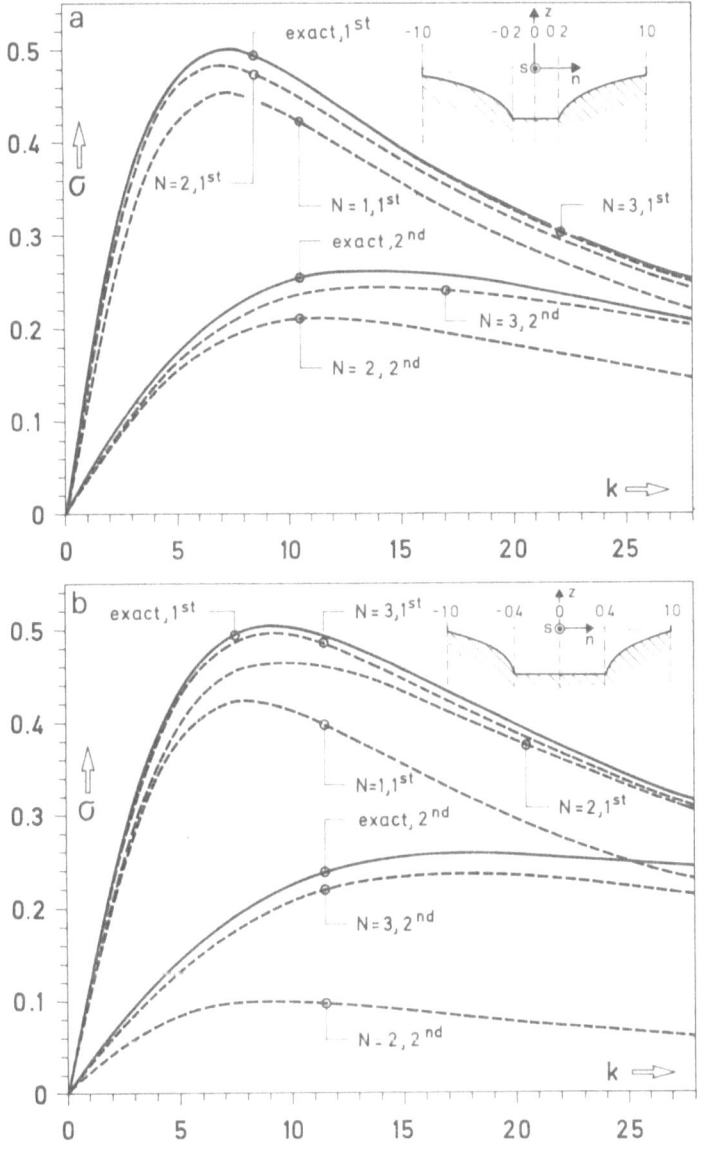

Figure 5.7   Comparison of the dispersion relation σ(k) of the exact
solutions in a piecewise exponential channel (see inset)
with the MWR solutions for N = 1, 2, 3 and the two first
modes.

(N = 1) and 5.8b (N = 3) indicates clearly how sensitively the dispersion
relation reacts to the topography. Generally, an increase of q shifts
the dispersion relation to smaller frequencies; thus periods at the
same wavenumber become longer. This could already be inferred from the
fact that topography gradients tend towards the boundary as q increases.

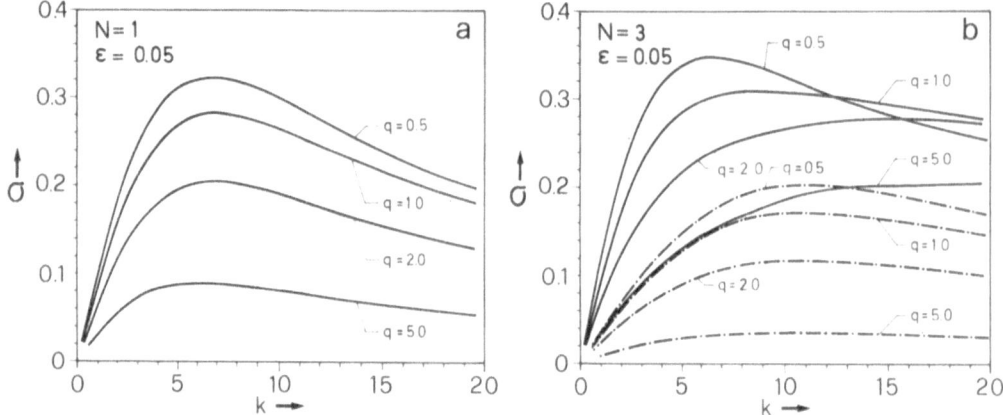

Figure 5.8   Effect of topography on the dispersion relation in a channel,
— first mode, —·—·— second mode, (a) N = 1, (b) N = 3.

Correspondingly, the slope parameter $|h'/h|$ grows which generally lowers
the frequencies. Comparison of the dispersion curves for the first mode
also indicates that the first order model may reproduce the dispersion
relation for strongly convex profiles (q = 0.5) quite adequately, while
it is definitely inadequate when profiles are triangular or concave.

Finally, *Figure 5.9* displays the dispersion relation of a second order
model for two different values of the sidewall parameter ε and for both,
convex and concave depth profiles. The latter are less affected by ε
than the former because all convex profiles of the form (5.1) join the
sidewall horizontally. The sidewall effect consists of a decrease of the

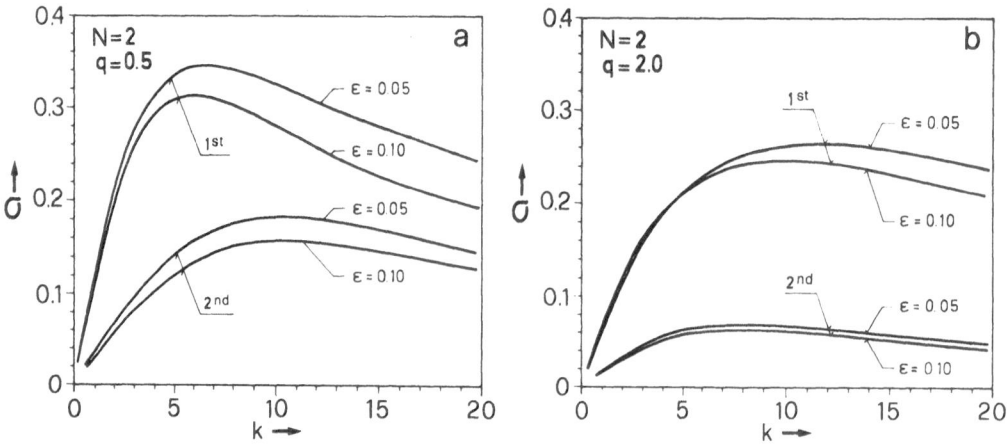

Figure 5.9   Effect of the sidewall parameter ε on a) convex (q = 0.5)
and b) concave (q = 2.0) profiles in a second-order model.

The question of whether k_0 at which $\partial\sigma/\partial k = 0$ exists for all topographies or wander off to infinity is of some practical significance. Figure 5.10 displays k_0 against the topography parameter q for a few values of ε. Whereas for convex profiles k_0 hardly depends on the sidewall parameter ε this is not so for concave profiles; a decrease of ε conspicuously increases the values of k_0. Alternatively, for large topography parameters k_0 is fairly independent of q. It is evident that models with very small sidewall parameters have very large critical wavenumbers. This problem is also considered in section 5.5, Figure 5.22.

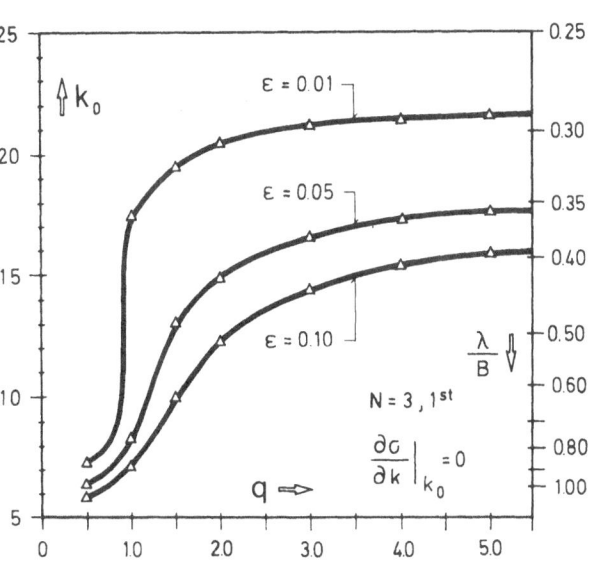

Figure 5.10

Plot of the critical wavenumber k_0, where the group velocity vanishes, as a function of topography and sidewall parameter for the first mode in a third-order model.

5.3 Channel solutions

Equation (5.8) represents a general solution in a straight, infinite channel with arbitrary cross-section. ψ is a complex-valued function and so, both real and imaginary parts are physically reasonable solutions. However, as can be easily shown, they differ only by a spatial or temporal phase shift. We recall the identities

$$\text{Im}(z) \equiv \text{Re}(-i\,z), \quad z \in \mathbb{C},$$

$$-i \equiv e^{-i\cdot\pi/2},$$

and obtain from (5.8)

$$\text{Im}(\psi(s,n,t)) = \text{Re}(e^{-i\cdot\pi/2}\,\psi(s,n,t))$$

$$= \text{Re}(\psi(s,n,t + T/4)).$$

Therefore, the complete information about the solution ψ is already obtained when considering $\text{Re}(\psi)$ alone.

Before discussing the solutions in detail, however, a qualitative argument is shown by which the stream function is related to the barotropic velocity field according to

$$\underline{u}^{bt} = \frac{1}{h} \, (\hat{\underline{z}} \times \nabla \psi) . \qquad (5.9)$$

It follows from this, that the deeper the channels are, the weaker the velocities will be. Further, convex stream function surfaces are connected with anti-cyclonic velocity cells (*Figure 5.11*), and the steeper the ψ-surfaces are the stronger will be the velocities in these cells.

Rather than considering general solutions such as (5.8) we investigate solutions to particular wavenumbers.

Figures 5.12-5.14 display perspective views and contour lines of $\text{Re}(\psi)$ in a straight infinite channel for a third order model. The pattern consists of two right-bounded topographic waves evolving from the superposition of the solutions $\psi(\sigma,k)$ and $\psi(\sigma,-k)$. Each mode shows its own characteristic cross-channel behavior. As would be expected, the complexity of the system of gyres increases with the mode number.

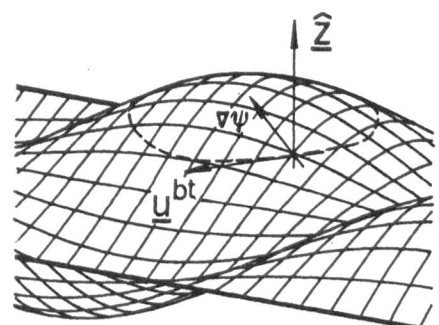

Figure 5.11

Explaining the anticyclonic barotropic velocity field on a convex stream function surface.

We now study the properties of the solution (5.8) for a single wavenumber. *Figure 5.15* exhibits the quality of the approximate solutions. Calculations have revealed that for a convex topography solutions converge rapidly for a wide range of wavenumbers, a result which is in accord with the observations above. For a concave topography ($q = 5.0$, *Figure 5.15*) the third-order solution is an acceptable approximation when $k = 2$ (*Figure 5.15a*); however, as *Figure 5.6* has already suggested, convergence for higher wavenumbers is slower (*Figure 5.15b*). Convergence is obviously also influenced by the choice of basis functions and it seems that the trigonometric functions are an appropriate set for small wavenumbers. It was a straightforward choice and made for analytical and computational simplicity. There may, however, be other complete sets, fulfilling the boundary conditions, which provide better results in some special cases. With the (sin, cos)-set the exact transverse functional dependence is well

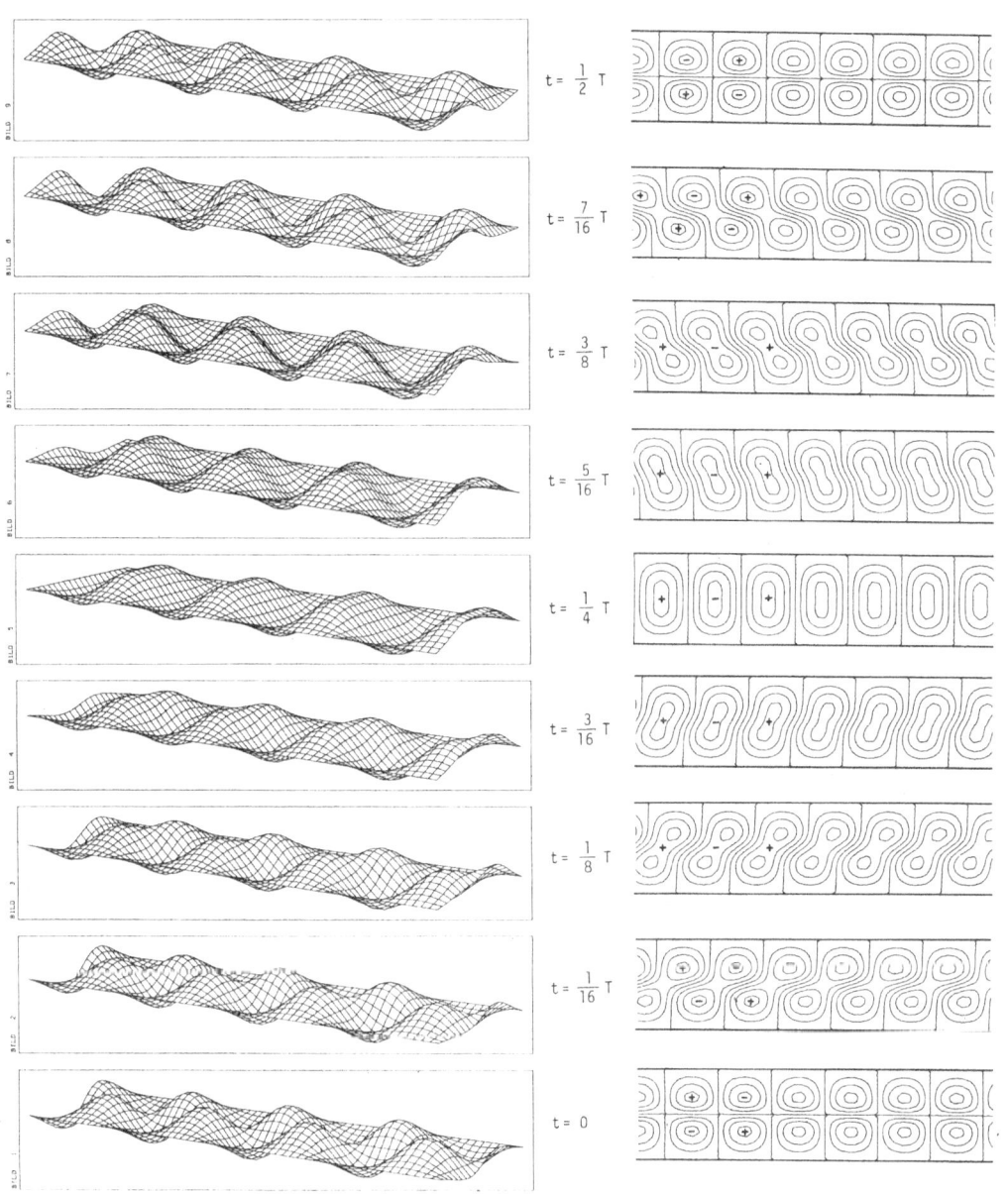

$N = 3, \quad \varepsilon = 0.05, \quad q = 0.5, \quad \sigma = 0.314, \quad T = 53.8h, \quad k = 4.00$

1st Mode Unit

Figure 5.12   a)   b)

Time sequence of the stream function surface in steps of $1/16\ T$ in a channel $-1/2\ B \leq n \leq 1/2\ B$, $0 \leq s \leq 6\ Lr$ and aspect ratio $r = 1$. Note that the phase motion in the domains $n > 0$ and $n < 0$ is right bounded.

Time sequence of lines of constant ψ relative to 90 % of the maximum value at each time step. The cellular structure of cyclonic (+) and anticyclonic (−) vortices is clearly visible.

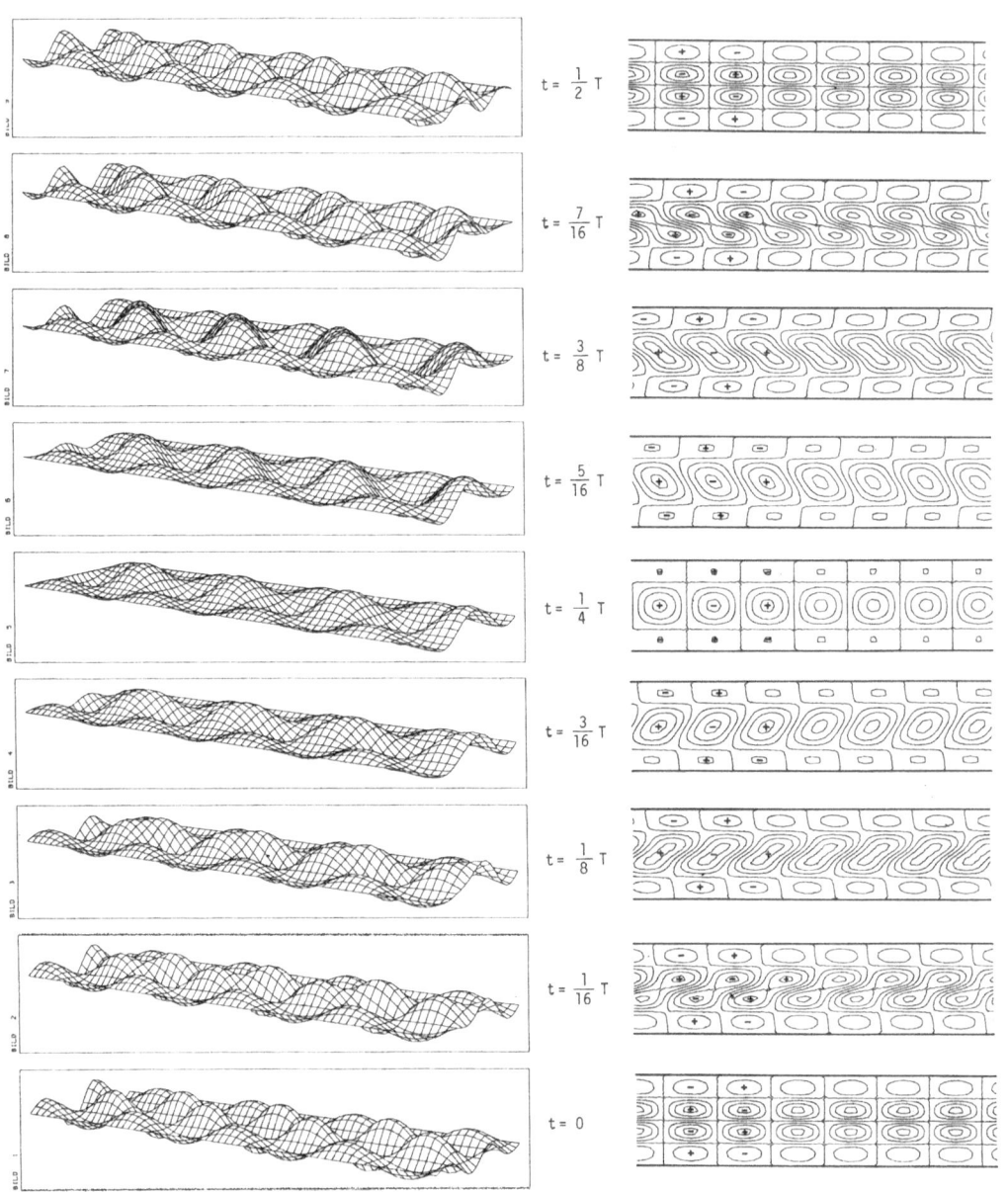

$$N = 3, \quad \varepsilon = 0.05, \quad q = 0.5, \quad \sigma = 0.127, \quad T = 133h, \quad k = 4.00$$

2nd Mode Unit

Figure 5.13   a)   b)

Time sequence of the stream function surface in steps of ¼16 T in a channel $-\sqrt{2}\ B \leq n \leq \sqrt{2}\ B$, $0 \leq s \leq 6\ Lr$ and aspect ratio $r = 1$. Note that the phase motion in the domains $n > 0$ and $n < 0$ is right bounded.

Time sequence of lines of constant ψ relative to 90 % of the maximum value at each time step. The cellular structure of cyclonic (+) and anticyclonic (-) vortices is clearly visible.

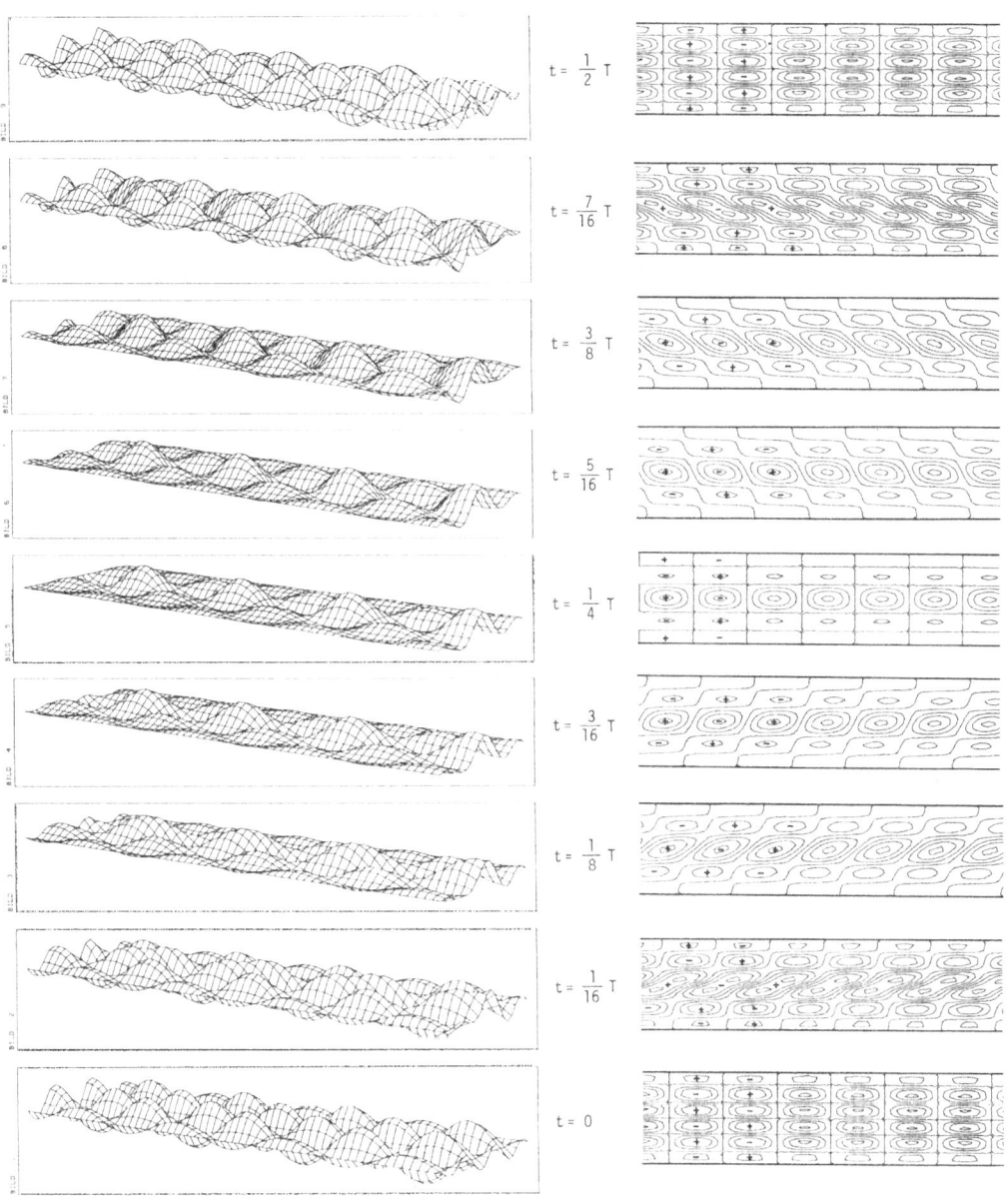

$$N = 3, \quad \varepsilon = 0.05, \quad q = 0.5, \quad \sigma = 0.0573, \quad T = 295h, \quad k = 4.00$$

3rd Mode Unit

Figure 5.14   a)   b)

Time sequence of the stream function surface in steps of $\sqrt{16}\ T$ in a channel $-\sqrt{2}\ B \le n \le \sqrt{2}\ B,\ 0 \le s \le 6\ Lr$ and aspect ratio $r = 1$. Note that the phase motion in the domains $n > 0$ and $n < 0$ is right bounded.

Time sequence of lines of constant ψ relative to 90 % of the maximum value at each time step. The cellular structure of cyclonic (+) and anticyclonic (-) vortices is clearly visible.

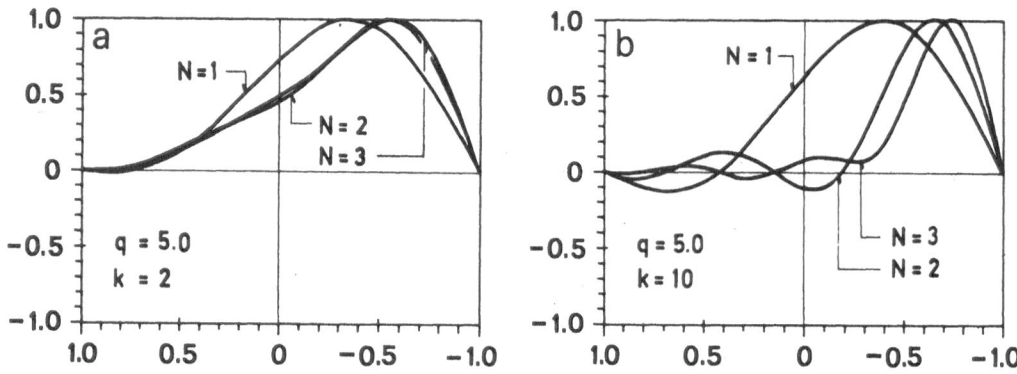

Figure 5.15
Convergence properties of the stream function of the first mode scaled
to a maximum value 1.0. The view is in the positive s-direction into
which the phase propagates in a right-bounded way. The sidewall parameter ε = 0.05 is selected, q = 5, (a) k = 2; (b) k = 10.

modelled for fundamental modes with not too large wavenumbers and small
topography parameters.

Figure 5.16 analyses the effect of the cross-sectional topography on the
stream function using q as a parameter. In view of the previous results,
a third-order model is anticipated to be sufficiently accurate. The effect for small wavenumbers (k = 2) and the first mode (*Figure 5.16a*) is comparatively weak; wave activity is slightly shifted towards the right
boundary for increasing topography parameters. Larger wavenumbers enhance
this effect.

For the second-mode solutions an increase of the topography parameter
again causes a shift of the ψ-surface towards the right boundary, see
Figure 5.16b. The right-most crest, however, is weakened and for larger topography parameters the main activity is in the middle crest.

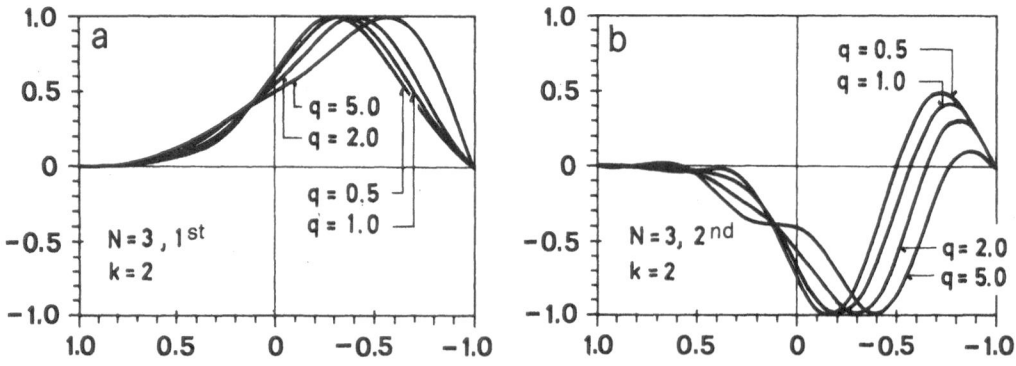

Figure 5.16
Transverse topography dependence of the stream function for the wavenumber k = 2 and the first two modes, N = 3, (a) first mode; (b) second mode.

Evidently, the transverse structure of topographic Rossby waves also depends strongly on the wavenumber k. This effect is comparable in magnitude with that of the topography. *Figure 5.17* demonstrates this for both types of topographies and the first two modes.

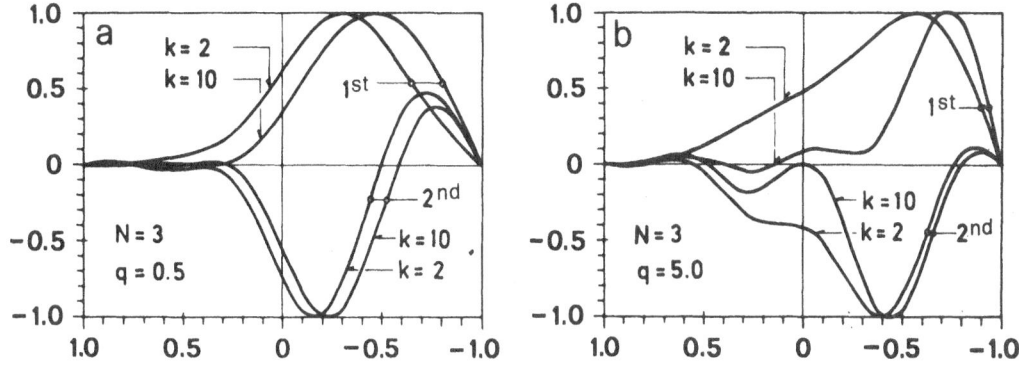

Figure 5.17
Wavenumber effect of the stream function for convex and concave topography, $\varepsilon = 0.05$ and the first two modes of a third-order model, (a) $q = 0.5$; (b) $q = 5.0$.

An increase of k generally shifts the stream function towards the right shoreline. The effect is large (small) for profiles with large (small) q particularly for the first-mode unit. Topography and wavenumber effect, therefore, act in the same way. These properties have not been clearly demonstrated in previous work. Suffice it to state that they have important practical bearings when mooring sites are projected.

5.4 Velocity profiles

The general channel solution (5.8) which satisfies the homogeneous system (4.9) is determined up to a constant factor. In order to compare different velocity profiles this constant should be fixed by using a further criterion. It seems reasonable to scale the occurring wave patterns by normalizing the free constant such that the global kinetic energy content is constant. (There is no potential energy for topographic Rossby waves in a rigid-lid formulation). Here, the problem is posed in terms of the barotropic mass transport stream function and a solution yields information about a *depth averaged* velocity field. This allows the calculation of only a *lower limit* of the true kinetic energy content.

The kinetic energy per unit mass that is contained in an infinitesimal volume is

$$d^3 E_{kin} = \frac{1}{2} (u^2 + v^2) \, J \, dn \, ds \, dz , \qquad (5.10)$$

in which the velocity components u, v can be expressed in terms of the stream function using (5.9), and for straight channels J = 1.

A minimum average energy density is obtained by integrating (5.10) across the channel axis and over the vertical, operating with

$$\overline{(\cdot)} = \lim_{T \to \infty} \frac{1}{T} \int_0^T (\cdot) \, dt, \quad \lim_{L \to \infty} \frac{1}{L} \int_0^L (\cdot) \, ds$$

and dividing by the cross-sectional area (T and L are not necessarily related to the spatial and temporal periods). It then reads (γ fixed)

$$\overline{E_{kin}} = \frac{1}{1+\varepsilon - \frac{1}{q+1}} \int_{-1}^{1} \frac{dy}{h} (k_\gamma^2 | P_\alpha^+ \, c_{\alpha\gamma} + P_{\alpha-N}^- \, c_{\alpha\gamma}|^2 + 4 | P_\alpha^{+'} \, c_{\alpha\gamma} + P_{\alpha-N}^{+'} \, c_{\alpha\gamma}|^2),$$

where y = 2n/B and ' = d/dy. When the stream function is scaled by $1/(\overline{E_{kin}})^{1/2}$ each wave contains the same kinetic energy. This enables comparison of the strength and structure of a wave pattern as a response to a given energy input.

Figure 5.18 displays the amplitude distributions of the alongshore and cross-channel velocity profiles for the first mode at k = 10 and ε = 0.05 for four different topography parameters q. Sign changes correspond to a phase shift of 180°. Evidently, the u-component indicates a strong right-bounded coastal jet which is well known in forced circulation models (Simons, 1980). Its strength depends upon the parameters q and ε. An increase of q lowers the absolute value of the velocity components considerably.

We have also observed, and *Figure 5.18* provides partial corroboration,

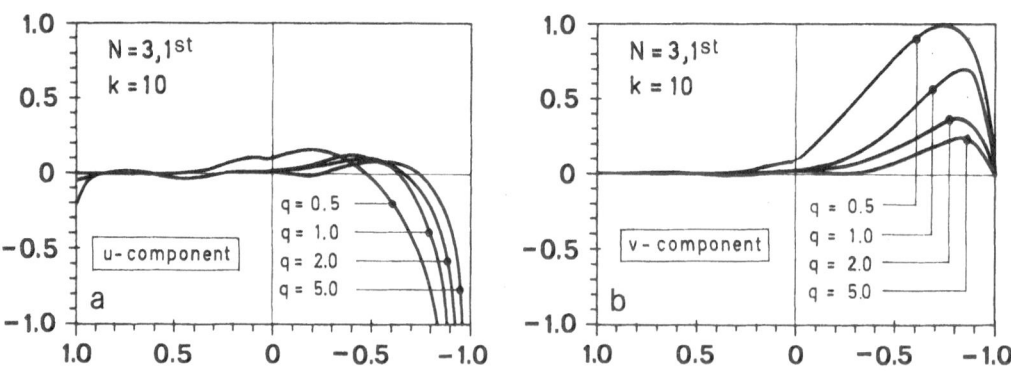

Figure 5.18

Transverse topography dependence of the depth-averaged velocity components (a) u (along-channel) and (b) v (across-channel) for N = 3, k = 10, ε = 0.05 and the first mode. All profiles are scaled such that the kinetic energy contents are comparable.

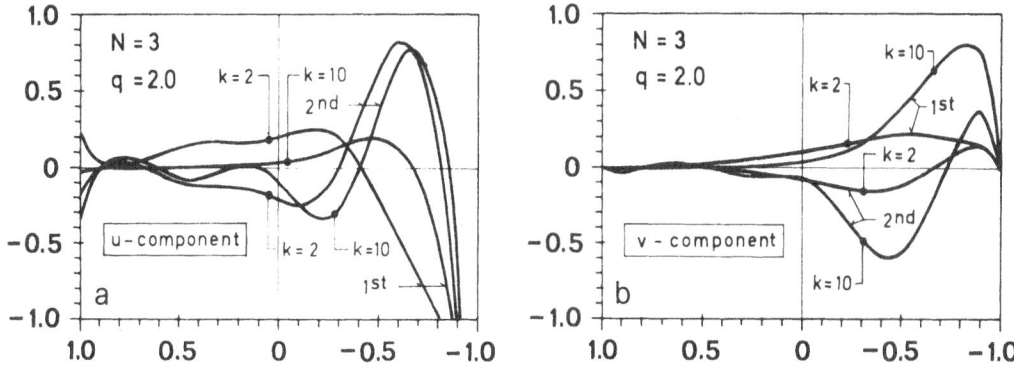

Figure 5.19
Wavenumber effect of the depth-averaged velocity components for q = 2.0,
ε = 0.05 and the first two modes of a third-order model, (a) u-component; (b) v-component.

that convergence of at least u is slower than that for the stream function. The reason is, of course, differentiation. Deviations of the computed velocity profiles from what they should be occur at the left shore-
line (*Figures 5.18a and 5.19a*).

Figure 5.19 illustrates the wavenumber effect for the case q = 2 (parabolic) and ε = 0.05. With growing wavenumber, activity in the u-component
shifts to the right shore and, correspondingly, activity diminishes in
the left part of the channel. Alternatively, cross-channel components
grow with increasing k. Therefore, long waves exhibit particle motion
which is mostly along the channel axis. Shorter waves with wavelengths
smaller than about a channel width have velocities of comparable order
in both directions. These properties also hold for the second mode.

As anticipated when introducing the sidewall parameter ε its effect on
the depth-averaged velocity profiles is very weak and only recogizable
in the u-component and close to the shoreline. *Figure 5.20* demonstrates
this for a channel with parabolic depth-profile. Velocity profiles dif-

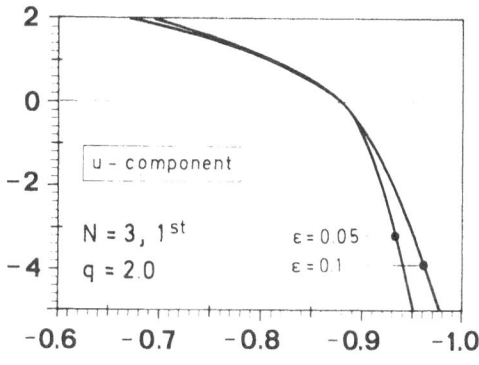

Figure 5.20

Effect of the sidewall parameter
ε on the u-compoent (along-channel) at k = 10 and with q = 2.0.
Because the profiles differ from
each other only at the right shore
this domain is enlarged, N = 3,
first mode.

fer from each other only very close to the right boundary. There, the
u-component of the velocity vector is directly governed by ε and its absolute value increases as ε approaches zero.

The above results can be used to answer questions which arise when topographic wave motion in channels or narrow elongated lakes is to be
detected and recorded. Scrutiny of the wavenumber dependence shows that,
in order to record the first mode on a concave topography ($q = 2.0$, $q =
5.0$), the mooring system is best placed within a domain that is 0.05 B
(B is the channel width) away from the shore. Then, both velocity components are of comparable magnitude and a whole range of wavenumbers
can be detected with a velocity vector which turns clockwise. The second
mode can most likely be detected within a domain which is 0.1 B to 0.2 B
away from the shore. For a proper test of the wave structure two moorings at the same side of the channel are desired.

5.5 Alternative solution procedures

Instead of applying the MWR to the TW-equation, one can start directly
from equation (2.22), introduce the plane-wave-trial solution

$$\psi(x,y) = F(y) \exp\left(i(kx - \sigma t)\right)$$

and deduce the two-point boundary value problem (TPBVP) for the transverse distribution function $F(y)$:

$$F'' - \frac{h'}{h} F' - \left(k^2 - \frac{h'}{h}\frac{k}{\sigma}\right) F = 0, \qquad y_1 < y < y_2,$$
$$F = 0, \qquad\qquad\qquad y = y_1, y_2. \tag{5.11}$$

Gratton (1983), Gratton & Le Blond (1986), Bäuerle (1986) and Bäuerle &
Hutter (1986) solve (5.11) for different channel topographies, whereas
Lie (1983), Djurfeldt (1984) and Takeda (1984) perform a shelf-wave ana-
lysis ($y_2 = \infty$) [*)].

Gratton and Le Blond (1986) investigate a channel with linear (asymmet-
ric) or parabolic depth profile and $y_1 = -y_2$. They discuss two types of
approximate solutions. In the first, the so called *small slope approximation*,
h is regarded as a constant except when differentiated. For large bottom slopes, both h and h' are treated as functions of y. They show that
for these profiles the solution of (5.11) can be expressed in terms of

---

*) Some of these authors formulate the problem in terms of the surface elevation
instead of the stream function. The emerging ODE is, however, similar to (5.11),
compare (3.1).

special functions[*]. In their configuration h'/h is bounded everywhere and so it is argued that Taylor series solutions about the interior regular point $y = 0$ are more convenient. This expansion consists of two linearly independent Taylor series, viz.

$$F = A \sum_{n=0}^{\infty} C_n \, y^n + B \sum_{n=0}^{\infty} D_n \, y^n. \qquad (5.12)$$

The coefficients are selected such that $C_0 = 1$, $C_1 = 0$ and $D_0 = 0$, $D_1 = 1$, and C_n, D_n, $n \geq 2$ are determined by substitution of (5.12) into (5.11). The free constants A, B follow from the boundary conditions $F(-y_2) = 0$ and $F(y_2) = 0$. The former yields

$$B = -A \frac{\sum C_n (-y_2)^n}{\sum D (-y_2)^n},$$

and the latter leads to the implicit *dispersion relation*

$$\left[\sum C_n \, y_2^n \right] \left[\sum D_n (-y_2)^n \right] - \left[\sum D_n \, y_2^n \right] \left[\sum C_n (-y_2)^n \right] = 0.$$

C_n and D_n depend on σ and k for $n \geq 2$. For the V-shaped channel the jump condition at $y = 0$ must also be accounted for.

Gratton and Le Blond show dispersion relations for the first three modes for both V-shaped and parabolic bathymetries which are qualitatively as those of *Figures* 5.6 - 5.9. They refrain, however, from discussing the numerical properties of this solution procedure (convergence, truncation of the series).

When h(y) is not so easily expressible in terms of analytic functions, solution of (5.11) by standard numerical techniques is probably more economical. Because of the nature of the TPBVP (5.11) the shooting method using either the Runge-Kutta method or any other high order multistep forward finite difference scheme that may account for the stiffness of the equation at large k or large h'/h may be the most efficient approach.

By solving the eigenvalue problem (5.11) with straightforward finite difference techniques, replacing derivatives with central second order difference expressions we will obtain a feeling and information about the reliability of numerical solutions of the TW-equation in two-dimensio-

---

[*] *The small slope approximations lead to elementary functions (linear profile) and to parabolic cylinder functions (parabolic bottom), the solutions of the full equations can be expressed in terms of Kummer functions (linear profile) and generalized spheroidal functions (parabola), respectively.*

nal domains. Bäuerle (1986) and Bäuerle & Hutter (1986) applied this technique to straight infinite channels and used the depth profile (5.1). It is instructive to compare their findings with the MWR results of sections 5.1 - 5.4. *Figure 5.21a* displays the dispersion relation of TW's in a parabolic channel for an increasing number NN of grid points. The side wall parameter ε is chosen very small in order to demonstrate the sensitivity of the numerical results with respect to ε. For larger values of ε convergence is generally better; fewer mesh points can be selected to obtain reliable results. We infer from *Figure 5.21a* that NN = 13 yields acceptable results only for very small wavenumbers. Due to the small value of ε the qualitatively correct behavior, i.e. very large critical wavenumber k_0 for which $\partial\sigma/\partial k = 0$, is predicted only when NN = 101 and larger. Moreover, convergence is decelerated for increasing q and k. *Figure 5.21b* illustrates this for the fundamental mode 1, q = 0.5, 2 and 9, and two wavenumbers k = 10, 30. Hence, high resolution is required to obtain satisfactory numerical prediction of the dispersion relation and mode structure, and this resolution must be higher for concave than convex profiles. This is unfortunate because concave profiles are more

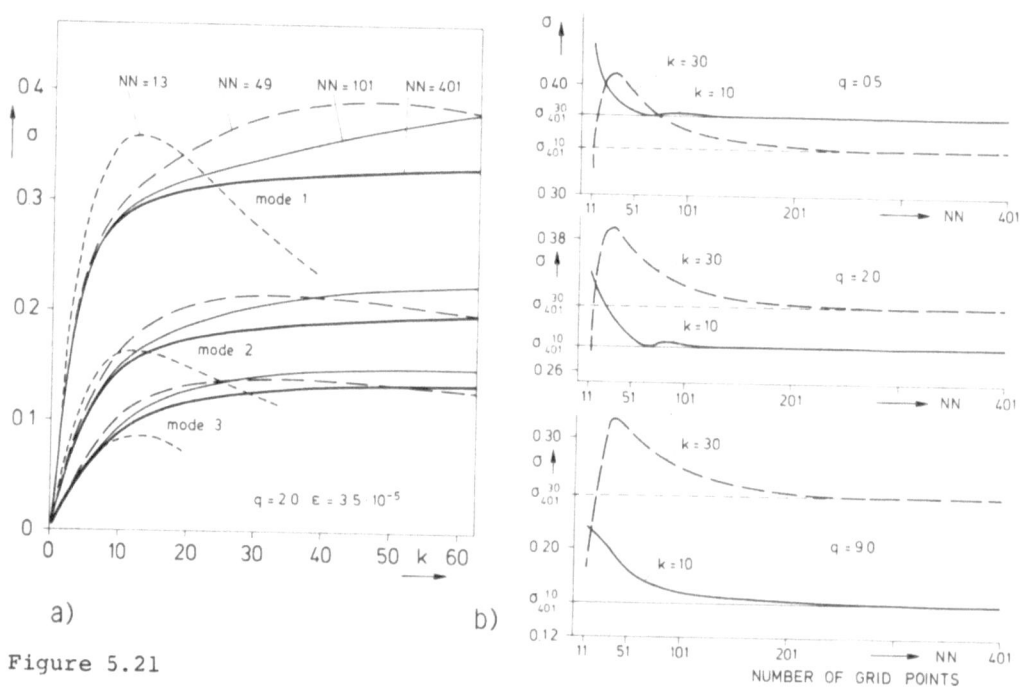

Figure 5.21

Convergence properties of the dispersion curves (a) and of the frequency σ(k) for the wavenumbers belonging to the first mode (b) for increasing number of grid points. σ_{401}^{10} and σ_{401}^{30} denote asymptotic frequencies for NN = 401 and k = 10, 30, respectively, and ε = 3.5·10⁻⁵. [From Bäuerle, 1986]

realistic. But the result also suggests caution with determined mode structures and periods of two-dimensional topographic waves in enclosed basins where coarser resolutions are necessary because of cost or memory limitations of the available computer device. Observations to this effect were made by Bennett & Schwab (1981).

Figure 5.22 corresponds to *Figure 5.10* and exhibits the sensitivity of the critical wavenumber k_0 with respect to the number of grid points NN. Differences of the curves $k_0(q)$ for NN = 49 and NN = 401 are observed mainly for small sidewall parameters ε. For these and for 49 grid points the dispersion relation leads to $k = k_0$ which is independent of q whenever q > 5; this is refuted when NN = 401 and a linear behavior emerges. Thus, one has to be very careful when selecting small sidewall parameters or, more generally when h'/h is large in the domain of integration.

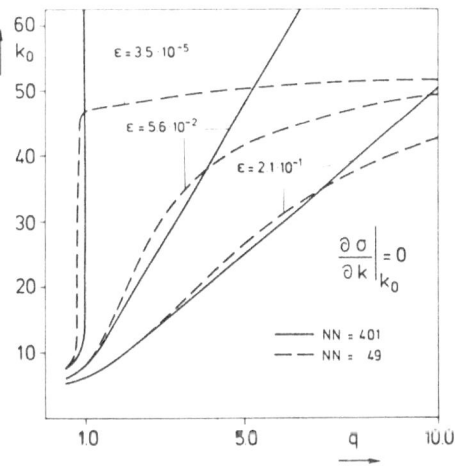

Figure 5.22

Critical wavenumber k_0 at which the group velocity vanishes plotted against the topography parameter q for three values of the sidewall parameter.
[From Bäuerle, 1986]

Bäuerle (1986) also compares the dispersion relations for ε = 0.05 and q = 0.5, 1, 2, 5 as determined by his finite difference technique (NN = 401) and the MWR using a third order model, see *Figure 5.23*. For q ≤ 2 and the indicated wavenumber range the curves agree satisfactorily; deviations are observed for large k and q. Such configurations require higher order MWR-models as has already been pointed out in section 5.2.

This approach clearly shows that excessive resolution of the channel width is necessary to achieve numerically reliable dispersion relations and mode structures. Because of the wave trapping schemes with variable mesh size or higher order finite differencing might, perhaps, be advantageous in lowering the total number of mesh points. However, the results indicate that one ought to be cautions with any coarse finite dif-

ference or finite element resolution in spatially two-dimensional domains. Elongated domains are therefore prone of requiring a large number
of grid points.

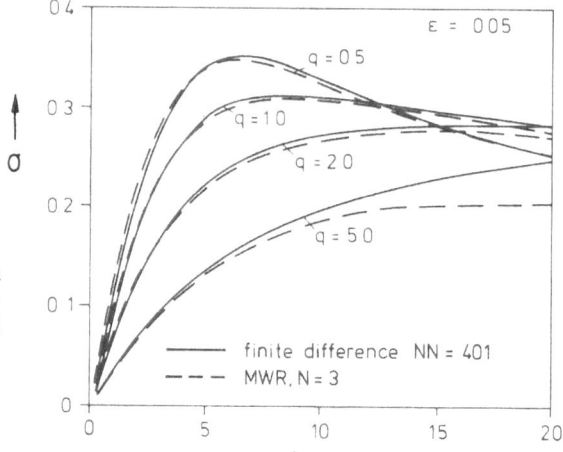

Figure 5.23

Comparison of the dispersion
curves calculated by the finite difference (solid) and
the MWR-technique (dashed).
[From Bäuerle, 1986]

This last remark may provide (heuristic) indications why our MWR-ap-
proach may be of advantage when one is attempting to solve the TW-equation in a two-dimensional elongated domain. Basically a relatively small
number of shape functions seems to guarantee a sufficiently accurate
resolution of the problem in the transverse direction. The problem in
the long direction becomes a vector ODE-equation, possibly with variable coefficient matrices. Thus, when the problem is solved in the long
direction high accuracy ODE-software can be used with advantage. We
will demonstrate this in Chapters 6 and 7.

5.6 Hyperbolically curved channels

We close this chapter by demonstrating how TW-solutions can be constructed in channels of which the shore lines follow confocal hyperbolas. Three such configurations where topographically trapped waves may
exist are illustrated in *Figure 5.24*. *Figure 5.24a* is a sketch of such a
wave along a curved boundary, the configuration in b) may be appropriate
to model TW's as they approach an isthmus or isolated basin and c) may
serve as a model of a curved channel. Provided that isobaths are confocal hyperbolas, i.e. the bathymetry only changes along the corresponding
confocal ellipses the previous solutions for straight channels can directly be used to find the dispersion relation and mode structures for
the TW's in these curved domains.

Consider the configuration of *Figure 5.25* in elliptical coordinates

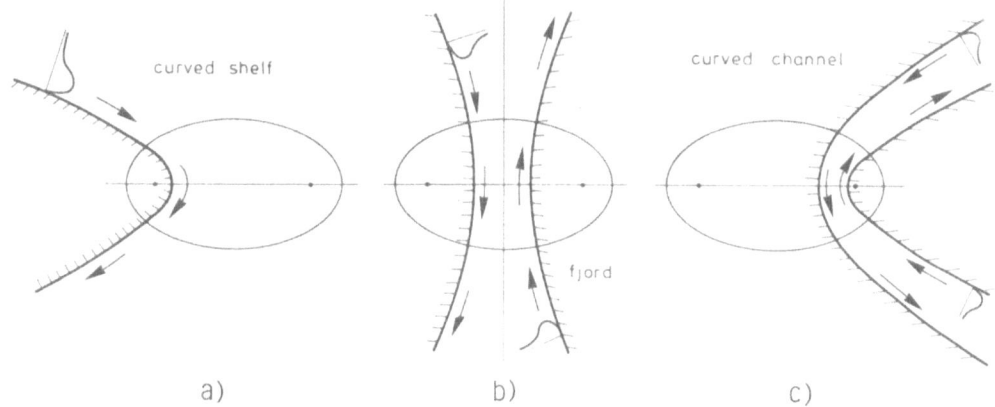

Figure 5.24   Sketch of the propagation of trapped waves along a hyperbolically curved shelf (a) or channel (b, c).

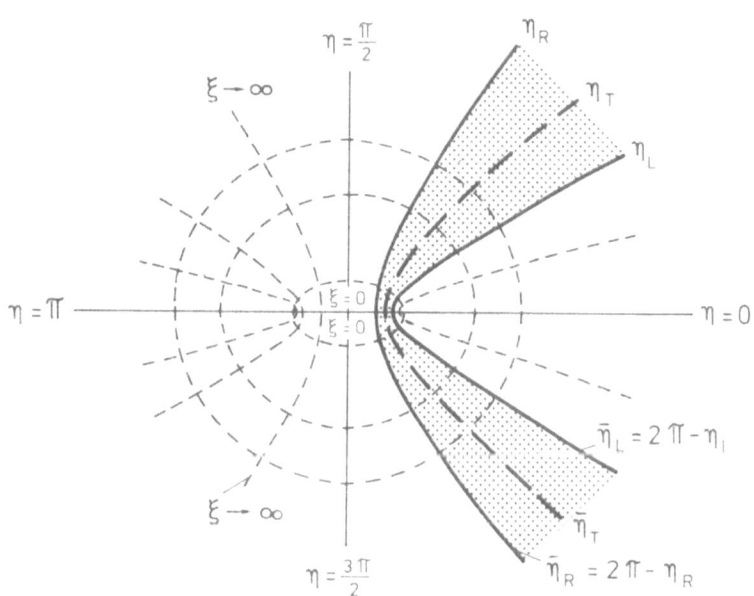

Figure 5.25
Hyperbolically curved channel embedded in an elliptical coordinate system with shore lines following confocal hyperbolas.

(ξ, η) and assume that in these coordinates $h = h(\eta)$. In this case the TW-
equation is given by (2.84). With the trial solution

$$\psi(\xi, \eta) = \begin{cases} F(\eta)\, e^{i(k\xi + \sigma t)}, & 0 \le \eta_L < \eta < \eta_R \le \pi, \\ G(\eta)\, e^{i(-k\xi + \sigma t)}, & \pi \le \bar{\eta}_R < \eta < \bar{\eta}_L \le 2\pi, \end{cases} \tag{5.13}$$

the following boundary value problems for F and G emerge

$$F'' - \frac{h'}{h} F' - (k^2 + \frac{h'}{h} \frac{k}{\sigma}) F = 0, \qquad \eta_L < \eta < \eta_R ,$$

$$F = 0, \qquad \eta = \eta_L, \eta_R ,$$

$$G'' - \frac{h'}{h} G' - (k^2 - \frac{h'}{h} \frac{k}{\sigma}) G = 0, \qquad 2\pi - \eta_R < \eta < 2\pi - \eta_L ,$$

$$G = 0. \qquad \eta = 2\pi - \eta_R, 2\pi - \eta_L .$$

(5.14)

Here primes mean differentiations with respect to η, and η_L and η_R denote the values of η of the hyperbolas along the boundary shore lines; they may have values from $0 \le \eta_L, \eta_R \le \pi$. Equation (5.13) represents a wave travelling in the hyperbolic channel approaching the narrowest cross section from above and leaving it in the lower half plane. The two waves match at $\xi = 0$ provided that

$$\psi(0,\eta) = \psi(0, 2\pi - \eta),$$

$$\frac{\partial \psi}{\partial \xi}(0,\eta) = - \frac{\partial \psi}{\partial \xi}(0, 2\pi - \eta), \qquad 0 \le \eta \le \pi$$

These conditions are fulfilled if

$$G(\eta) = F(2\pi - \eta).$$

This is consistent with (5.14)$_{2,4}$. We now demonstrate that (5.14) is formally analogous to the TPBVP in a *straight* channel. In order to fully apply the correspondence we transform the independent coordinates such that in the transformed coordinate the domain is the same as that for which the straight channel solution has been determined. *Table 5.2* demonstrates the different coordinate domains and the transformations. The TPBVP for the scaled Cartesian and the scaled elliptical system read [*)]

Cartesian	Scaled		Elliptic
$0 \le s \le L$ $\longrightarrow$	$0 \le \rho \le 1$ $\longleftarrow$		$0 \le \xi \le \xi_E$
$\rho = s/L$	$\rho = \xi/\xi_E$		
$-\frac{1}{2} B \le n \le \frac{1}{2} B$ $\longrightarrow$	$-\frac{1}{2} \le \chi \le \frac{1}{2}$ $\longleftarrow$		$\eta_L \le \eta \le \eta_R$
$\chi = n/B$	$\chi = \dfrac{\eta}{\eta_R - \eta_L} - \dfrac{\eta_R + \eta_L}{2(\eta_R - \eta_L)}$		

Table 5.2   Coordinate domains and transformations for the two systems.

[*)] *An identical equation as that for F can also be obtained for G if*

$$\chi = - \frac{1}{\eta_R - \eta_L} \eta - \frac{\eta_R + \eta_L}{2(\eta_R - \eta_L)} + \frac{2\pi}{\eta_R - \eta_L}$$

is chosen.

$$F_s'' - (\frac{h'}{h}) F_s' - \left[\frac{B^2}{L^2} k_s^2 + \frac{B}{L} (\frac{h'}{h}) \frac{k_s}{\sigma} \right] F_s = 0,$$

$$F_c'' - (\frac{h'}{h}) F_c' - \left[\frac{(\eta_R - \eta_L)^2}{\xi_E^2} k_c^2 + \frac{(\eta_R - \eta_L)}{\xi_E} (\frac{h'}{h}) \frac{k_c}{\sigma} \right] F_c = 0,$$

in which primes denote differentiations with respect to χ, and the subscripts s and c stand for "straight" (Cartesian) and "curved" (elliptic). These two problems are formulated in the same domain $[0 \leq \rho \leq 1, -\frac{1}{2} \leq \chi \leq \frac{1}{2}]$ and are identical if

$$k_s = \frac{(\eta_R - \eta_L)/\xi_E}{B/L} \cdot k_c . \qquad (5.15)$$

The effect of curvature is therefore measured by the two aspect ratios, $(\eta_R - \eta_L)/\xi_E$ and B/L in the two coordinate systems. In order to determine $\sigma(k_c)$ it thus suffices to stretch the k-axis of the straight channel dispersion relation according to (5.15). Likewise, the eigenfunctions are obtained from the straight channel solutions by a stretching transformation. Notice that this approach incorporates an entire family of curved channel solutions. For instance $\eta_R = \pi - \eta_L$, $0 < \eta < \pi/2$ may be appropriate to model TW's close to the mouth of a fjord, while $\pi/2 < \eta_R < \eta_L < 0$ is appropriate to model curvature effects in channels. Along the thalweg hyperbola

$$\eta \equiv \eta_T = \frac{\eta_R + \eta_L}{2}$$

this curvature is given by

$$K(\xi) = \frac{-\cos \eta_T \sin \eta_T}{a(\cos^2 \eta_T \sinh^2 \xi + \sin^2 \eta_T \cosh^2 \xi)^{3/2}}$$

and its maximum is obtained for $\xi = 0$

$$K(0) = - \frac{\cos \eta_T}{a \sin^2 \eta_T} .$$

The behavior of TW's in hyperbolic channels, however, depends on two parameters, the curvature K and a width parameter $\eta_R - \eta_L$.

6. Topographic waves in rectangular basins

As we learnt in chapter 3 there exist a number of analytical models for TW's in enclosed domains. Typical properties of TW-motion in these finite domains were found; among these were the conspicuous structure of the lowest modes (linear and quadratic Ball-mode), the counterclockwise phase propagation, the rotation of the velocity vectors, to name a few. In special cases these models allowed satisfactory interpretation of long-periodic phenomena in lakes. However, only a restricted number of parameters was offered to model a particular basin geometry. Moreover, observational results in Lake of Lugano and Lake of Zurich are still awaiting interpretation by models which account more accurately for their elongated shape, topography and, perhaps, include curvature effects. Particularly, the study of TW's in the Northern Lake of Lugano raised further questions regarding the applicability of the analytical models to this basin. On the one hand, the elliptical model of Mysak et al. (1985) could explain the 74h-trace in the measurements; it was interpreted as a (1,1)-TW-mode with a *global* wave pattern. On the other hand, Trösch (1984) applied a finite element method to the elongated basin and found a completely different wave behavior in the 65-100 h interval: Rather than a global pattern he observed *local* wave motion for which wave activity was

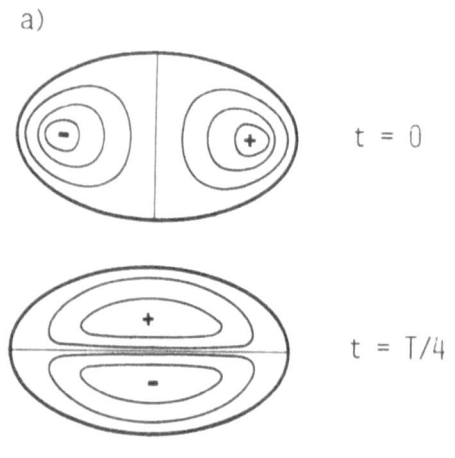

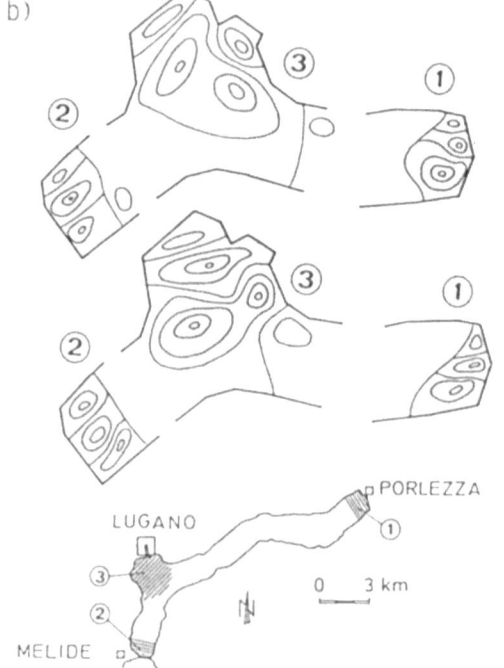

Figure 6.1

Are the results obtained by an exact model, a), and by applying a finite element technique to a realistic bathymetry, b), contradictory? See also *Figures 1.9 and 1.10*.

trapped to the three bays of Lake of Lugano, leaving the other parts of the lake calm[*]. *Figure 6.1* illustrates this apparent controversy. So we must ask the question: are exact models applicable to real basins and how can the finite element results be *physically* interpreted?

These open questions have motivated us to try to use the channel method developed in the previous chapter for the construction of a model of TW's in rectangular basins. Although the MWR together with the formulation of (2.22) in a natural coordinate frame offers the possibility of account- ing for curvature effects these are set aside here because we first wish to test the quality of such quasi-one-dimensional models. What follows is based on Stocker & Hutter (1985, 1986a,b).

6.1 Crude lake model

We call a lake model "crude" if in the natural coordinates (s,n) its topo- graphy varies only in the transverse direction n. For such a model it is straightforward to extend the results obtained for infinite channels. We use the depth profile (5.1) with $h_0(s) = \text{const}$. As there exist 4N in- dependent channel solutions of the form (5.5) in a Nth order model, these can be superposed to a lake solution. A crude lake model is obtained by inserting *vertical* walls at two positions $s = 0$ and $s = L$. At these points the stream function ψ must vanish. So, in view of (4.9) we have $\psi_\alpha = 0$ $(\alpha = 1, 2, \ldots, 2N)$ for $s = 0$ and $s = L$ and hence

$$\sum_{\gamma=1}^{4N} c_{\alpha\gamma}\, d_\gamma = 0 , \qquad (6.1a)$$

$$\left.\begin{array}{l} \\ \\ \end{array}\right\} \quad (\alpha = 1, \ldots, 2N) .$$

$$\sum_{\gamma=1}^{4N} e^{ik\gamma} c_{\alpha\gamma}\, d_\gamma = 0 , \qquad (6.1b)$$

Recall from chapter 5 that the coefficients $c_{\alpha\gamma}$ are functions of σ. This homogeneous system has a non-trivial solution provided that its determi- nant is zero. This selects the eigenfrequencies of the system which de- pend on the bathymetry given by r, q and ε. Periodic lake solutions have been seen to exist only for $0 < \sigma < \sigma_0$, where σ_0 denotes the maximum of the real branch of the first-mode unit. Consequently, frequencies de- crease appreciably when the topography parameter q increases. This ef- fect is demonstrated in *Table 6.1* which compares the first eigenfrequen- cies for models of different order. For a parabolic depth profile a

---

[*] *Local TW-motion was also observed in the numerical model of Lakes Ontario and Superior by Rao & Schwab (1976).*

2:1 basin	N = 1	N = 2	N = 3
q = 0.5	0.314	0.335	0.337
	0.292	0.316	0.317
	0.264	0.293	0.295
q = 2.0	0.198	0.260	0.274
	0.186	0.254	0.271
	0.169	0.246	0.267
q = 5.0	0.087	0.167	0.208
	0.081	0.163	0.206
	0.073	0.158	0.202

Table 6.1

First eigenfrequencies σ for $r = 0.5$ and $\varepsilon = 0.05$ in a simple lake model. There is always a pair of eigenfrequencies differing from each other by less than 1 % and the table shows only one of them. $N = 1, 2, 3$ indicates the order of the model.

q	r = 0.5	r = 0.4	r = 0.3	r = 0.2
0.5	0.335	0.337	0.339	0.341
1.0	0.303	0.304	0.304	0.305 ?
2.0	0.260	0.260	0.261	0.261
5.0	0.167	0.167	0.167	0.168 ?

Table 6.2

The first eigenfrequency in a second-order model for various aspect ratios r and topography parameters q, $\varepsilon = 0.05$. Question marks indicate computational difficulties.

third-order model offers adequate estimates of the eigenfrequencies. A parameter study also reveals that the topography parameter q influences the eigenfrequencies much more than do r or even ε, see *Table 6.2*.

Calculations further showed that for small aspect ratios system (6.1) is very difficult to handle. The smaller r is, the larger will be all $|Im(k)|$, and terms of (6.1b) become dominant; the smallest inaccuracies in the eigenvector d_γ are fatal because of their amplification in the terms proportional to $e^{ik\gamma}$. A remediable approach might be a superposition of two semi-channel solutions which are displaced with respect to each other by a length L.

Figure 6.2 shows a series of isolines of the stream function of a lake solution in a basin with a side ratio 2:1. The influence of the vertical walls is obvious in that wave crests approaching them die out. The fundamental mode does not resemble Ball-type behavior; rather, the wave patterns exhibit local structure. As the eigenfrequencies decrease the local character becomes stronger, but there is still a right-bounded phase propagation. *Figure 6.3* presents some specific lake solutions for other aspect ratios and for N = 2 and N = 3 models. The mode in *Figure 6.3a* is similar to a compound channel solution, wave patterns along the two opposite shores seem not to interact, whereas for a higher mode (*Figure 6.3b*) flow across the channel is observed. *Figure 6.3c* displays the stream function pattern of a very complex solution with strong local structure.

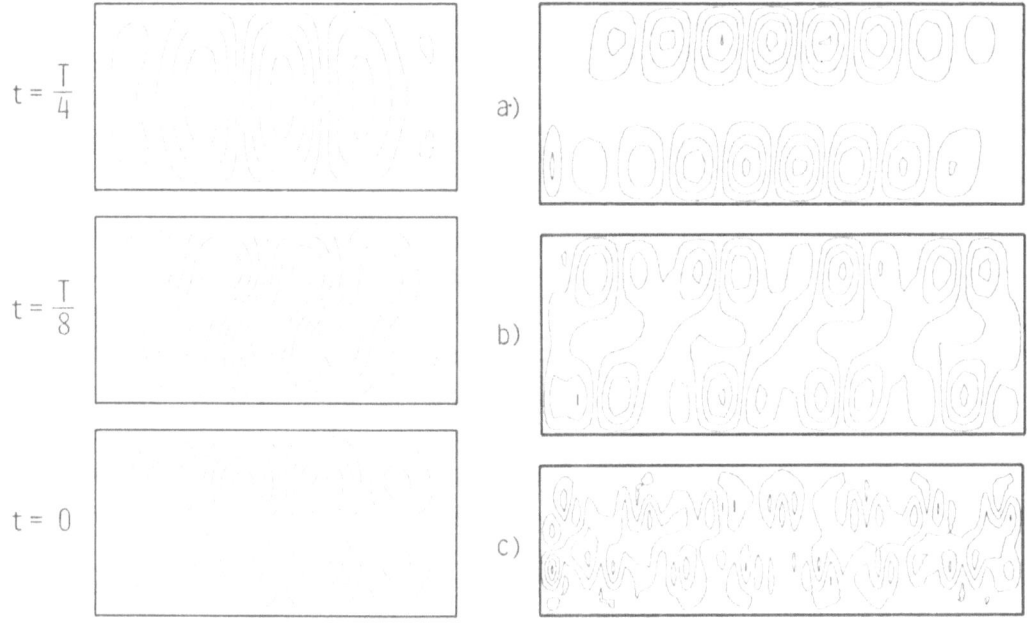

$t = \dfrac{T}{4}$

$t = \dfrac{T}{8}$

$t = 0$

a)

b)

c)

Figure 6.2

Lake solution in a 2:1 basin plotted for three different times through a quarter of a period T, using N = 2, q = 0.5, σ = 0.335. Wave activity is strongest in the middle of the basin and damped at both ends

Figure 6.3

Stream function of three examples of solutions in a crude lake model. The parameters are

	N	r	q	ε	σ
a)	2	0.4	2.0	0.05	0.260
b)	2	0.4	2.0	0.05	0.244
c)	3	0.3	0.5	0.05	0.120

Note that the basin center may not be an exact center of point symmetry. This is due to numerical inaccuracies and the different properties of the lake boundaries at s = 0 and s = L. The dependence on the wavenumber k_γ enters the boundary conditions (6.1a and b) differently. To no surprise, the asymmetry is particularly visible in *Figure 6.3c*. Choosing the coordinate s symmetrically would eliminate this imbalance.

These and many more results that were obtained are rather distressing and show no relation to the solutions of exact models. Very simple mode structures with a global phase propagation could not be found. This, however, is not surprising. The rectangular basin with vertical end-walls has lines of constant f/H (isotrophes, isobaths) which are *not* continuous: they start at one wall and reach the opposite wall in a straight line. It is known that the phase propagation tends to follow these lines and therefore, global phase motion around the basin (as in the linear Ball-mode) may not be expected here. Hence, there is a need for an improved lake model which has *continuous* depth lines.

Parenthetically we also remark that, in order to preserve the *algebraic* procedure (6.1) for the determination of the eigenfrequencies, a modified lake basin can be studied. This consists of three sections each of which exhibits an exponential depth profile of the form exp(-cs), see *Figure 6.4*. Three finite lake solutions can be patched together at the slope discontinuities $s = s_1$, $s = s_2$ and a system much like (6.1), but three times larger must be solved. We have done this for several configurations shown in *Figure 6.4*, however the results are no more encouraging. A likely reason is that, even though the isobaths now are continuous they are not differentiable at the intersection points s_1 and s_2. We therefore are urged to relax the assumption $h_s/h =$ constant and allow for an arbitrary $h(s,n)$ leading to smooth isobaths. In this case, the matrix operator $\underset{\sim}{\mathbb{K}}$ then has no longer constant coefficients.

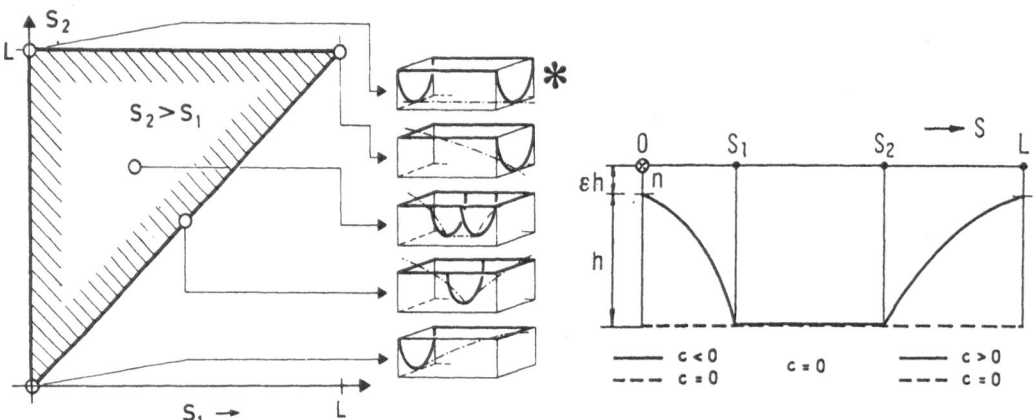

Figure 6.4
Variety of basin topographies in the parameter space (s_1,s_2) which can be treated after the refinement of the topography assumptions. In this section only solutions to topographies marked with ✳ were calculated.

6.2 Lake model with non-constant thalweg

a) Numerical method

Consider a rectangle of width B and length L which has the depth profile

$$h(s,n) = h_0(s)\left(1+\varepsilon-\left|\tfrac{2n}{B}\right|^q\right), \qquad 0 \le s \le L, \quad -\tfrac{1}{2}B \le n \le \tfrac{1}{2}B, \qquad (6.2)$$

with constant ε and $0 < q < \infty$. This bathymetry possesses a finite shore depth $\varepsilon h_0(s)$, which is necessary to have $(\partial h/\partial n)/h$ bounded everywhere. It was demonstrated in chapter 5 how the boundary value problem (2.24) was transformed to a new *one-dimensional* problem for the coefficient functions $\psi_\alpha^\pm(s)$. The result was

$$\begin{aligned} \underset{\sim}{\mathbb{K}}\,\underset{\sim}{\psi}(s) &= 0, & 0 < s < L, \\ \underset{\sim}{\psi}(s) &= 0, & s = 0,L, \end{aligned} \qquad (6.3)$$

in which

$$\underset{\sim}{\psi} = (\psi_1^+, \ldots, \psi_N^+; \psi_1^-, \ldots, \psi_N^-) = (\underset{\sim}{\psi}^+; \underset{\sim}{\psi}^-),$$

$$\begin{aligned}
\mathbb{IK} = &- i\sigma \left[B^2 \underset{\sim}{K}^{00} \frac{d^2}{ds^2} - B^2 (h^{-1} \frac{dh}{ds}) \underset{\sim}{K}^{00} \frac{d}{ds} - \underset{\sim}{K}^{22} \right] \\
&- B(\underset{\sim}{K}^{20} + \underset{\sim}{K}^{02}) \frac{d}{ds} + B(h^{-1} \frac{dh}{ds}) \underset{\sim}{K}^{20}.
\end{aligned} \tag{6.4}$$

Here and henceforth $h = h_0$, and it has been assumed that the operator $\mathbb{IK}$ has coefficients which depend on the variable s through an arbitrary thalweg depth $h(s)$. Furthermore, the symmetrized form of $\mathbb{IK}$ is obtained by using (5.3) to express the $\underset{\sim}{K}$'s in symmetrized form. For the numerical solution we transform (6.3) to a *real, first-order* system. Introducing

$$\underset{\sim}{\psi} \equiv (\text{Re }\underset{\sim}{\psi}^+, \text{Re }\underset{\sim}{\psi}^-, \text{Re }\underset{\sim}{\dot\psi}^+, \text{Re }\underset{\sim}{\dot\psi}^-, \text{Im }\underset{\sim}{\psi}^+, \text{Im }\underset{\sim}{\psi}^-, \text{Im }\underset{\sim}{\dot\psi}^+, \text{Im }\underset{\sim}{\dot\psi}^-), \tag{6.5}$$

with $(\ )^{\cdot} \equiv d/ds$ and substituting $s' \equiv s/L$, $d/ds' = Ld/ds$ we obtain after dropping primes

$$\frac{d}{ds} \underset{\sim}{\psi} = \underset{\sim}{A}(s) \underset{\sim}{\psi}, \qquad 0 < s < 1, \tag{6.6a}$$

$$B \underset{\sim}{\psi} = 0, \qquad s = 0,1. \tag{6.6b}$$

This system has dimension $8N$; $\underset{\sim}{B}$ is a constant diagonal matrix with $B_{ii} = 1$ for $i = 1, \ldots, 2N$ and $i = 4N+1, \ldots, 6N$ and else $B_{ii} = 0$. The matrix $\underset{\sim}{A}$ can be split up into a part which is independent of s and another part proportional to the slope parameter S

$$S(s) \equiv h^{-1} \frac{dh}{ds}, \tag{6.7}$$

explicitly

$$\underset{\sim}{A}(s) = \underset{\sim}{C} + S(s) \cdot \underset{\sim}{D}.$$

The matrices $\underset{\sim}{C}$ and $\underset{\sim}{D}$ take the form (the subscripts R and I stand for real and imaginary parts)

$$\underset{\sim}{C} = \begin{bmatrix} \underset{\sim}{C}_R & -\underset{\sim}{C}_I \\ \underset{\sim}{C}_I & \underset{\sim}{C}_R \end{bmatrix}, \qquad \underset{\sim}{D} = \begin{bmatrix} \underset{\sim}{D}_R & -\underset{\sim}{D}_I \\ \underset{\sim}{D}_I & \underset{\sim}{D}_R \end{bmatrix} \tag{6.8}$$

with the $(4N \times 4N)$-submatrices

$$\underset{\sim}{C}_R = \begin{bmatrix} 0 & 1 \\ \frac{1}{r^2}(\underset{\sim}{K}^{00})^{-1}\underset{\sim}{K}^{22} & 0 \end{bmatrix}, \quad \underset{\sim}{C}_I = \frac{1}{\sigma}\begin{bmatrix} 0 & 0 \\ 0 & \frac{1}{r}(\underset{\sim}{K}^{00})^{-1}(\underset{\sim}{K}^{20} + \underset{\sim}{K}^{02}) \end{bmatrix},$$

$$\underset{\sim}{D}_R = \begin{bmatrix} 0 & 0 \\ 0 & 1 \end{bmatrix}, \quad \underset{\sim}{D}_I = \frac{1}{\sigma}\begin{bmatrix} 0 & 0 \\ -\frac{1}{r}(\underset{\sim}{K}^{00})^{-1}\underset{\sim}{K}^{20} & 0 \end{bmatrix}, \tag{6.9}$$

and the aspect ratio $r = B/L$. The matrices (6.9) are independent of s and need be calculated only once during the integration for $s \in [0,1]$.

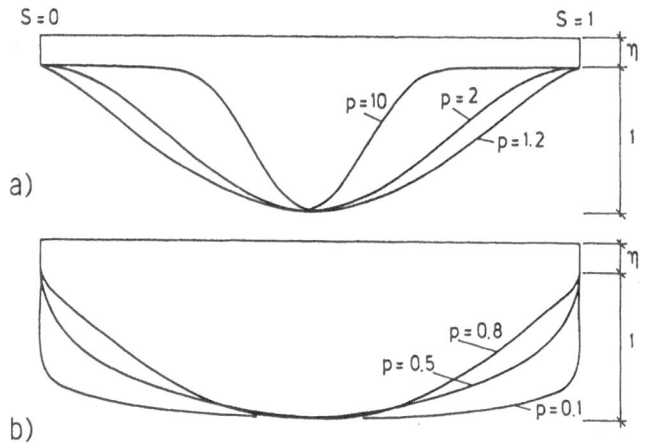

Figure 6.5

Thalweg profiles (6.10) for different values of the exponent p. For p>1 slopes at the lake ends are zero, a), when p<1 they are infinite, b).

Solutions of the TPBVP (6.6) were constructed numerically for the profile (*Figure 6.5*)

$$h(s) = \eta + \sin^P(\pi s), \tag{6.10}$$

here with p = 2. η and p are parameters; $\eta > 0$ guarantees that the depth is never zero and the exponent p could be varied such that the longitudinal variation of the depth is more or less concentrated at the long ends of the lake. The slope parameter S(s) is easily calculated from (6.10); one obtains

$$S(s) = \frac{p\pi \sin^{p-1}(\pi s) \cos(\pi s)}{\eta + \sin^P(\pi s)}. \tag{6.11}$$

For p > 1 and $\eta > 0$, S(s) vanishes at the lake ends, which is a numerical advantage, when 0 < p < 1 the slope parameter is not finite at s = 0,1. In order to keep S(s) finite everywhere (6.10) could be replaced by

$$h(s) = \begin{cases} \eta + bs, & 0 \le s \le \hat{s}, \\ \sin^p(\pi s), & \hat{s} \le s \le 1 - \hat{s}, \\ \eta + b(1-s), & 1 - \hat{s} \le s \le 1, \end{cases}$$

in which, for a given shore-slope b, η and $\hat{s}$ can be calculated such that h and h' are continuous at s = $\hat{s}$. With this choice (6.7) is finite everywhere and for all p > 0. The lake model now consists of two-side-wall parameters ϵ and η (or alternatively ϵ and the shore-slope b) and a longitudinal and transverse topography parameter p and q, respectively.

Equation (6.6a) allows the formal integration

$$\underset{\sim}{\Psi}(s) = \exp(\int_0^s A(\hat{s}) \, d\hat{s}) \, \underset{\sim}{\Psi}(0),$$

$$\equiv E(s) \, \underset{\sim}{\Psi}(0). \tag{6.12}$$

(6.6b) implies

$$\Psi(0) = (0, \dot{\psi}_R(0); 0, \dot{\psi}_I(0)), \tag{6.13a}$$

$$\Psi(1) = (0, \dot{\psi}_R(1); 0, \dot{\psi}_I(1)), \tag{6.13b}$$

and the symmetrization has been dropped for convenience of ensuing arguments. Formally, $E(s)$ in (6.12) is a matrix valued function. At the basin end, it can be written as

$$\left. E(1) \right|_{s=1} = \begin{bmatrix} E_{11} & \cdots & E_{14} \\ \cdot & & \cdot \\ \cdot & & \cdot \\ E_{41} & \cdots & E_{44} \end{bmatrix}. \tag{6.14}$$

Note that $E(1)$ is a function of the frequency σ via (6.12) and (6.9) and the E_{ij} are $(2N \times 2N)$-matrices. For each initial vector of the form

$$\Psi_j(0) = (\underbrace{0,0,\ldots,0}_{(j-1)},1,\underbrace{0,0,\ldots,0}_{(8N-j)}), \quad \begin{matrix} 2N + 1 \leq j \leq 4N, \\ 6N + 1 \leq j \leq 8N, \end{matrix} \tag{6.15}$$

the corresponding vector $\Psi_j(1)$ is computed using a discretized form of (6.6), see below. From (6.12) and (6.13) it then easily follows that the solution $\Psi_j(1)$ corresponding to the j-th initial vector $\Psi_j(0)$ is the j-th column of the matrix $E(1)$. (6.13b) eventually requires

$$\begin{bmatrix} E_{12} & E_{14} \\ E_{32} & E_{34} \end{bmatrix} \begin{bmatrix} \dot{\psi}_R(0) \\ \dot{\psi}_I(0) \end{bmatrix} = 0,$$

which allows derivation of the equation which determines the *eigenfrequency* in this lake basin. It takes the form

$$\det \begin{bmatrix} E_{12} & E_{14} \\ E_{32} & E_{34} \end{bmatrix} = 0. \tag{6.16}$$

It remains to select the integration routine for the 4N initial-value problems (6.6a) with (6.15). This choice depends on how the matrix $A(s)$ is available. Here A can be computed for all s $[0,1]$ and the fourth order *Runge-Kutta* scheme (or higher order multi-step schemes may be appropriate. This is a well known single-step forward integration technique[*]. We discretize the integration interval $[0,1]$ into M equidistant increments of length $d = 1/M$. The Ψ_{i+1} at the position s_{i+1} within the interval is then given by

---

[*] *The fundamental single-step forward integrator is the Euler-Cauchy scheme. It reads $\psi_{i+1} = \psi_i + dA(s_i)\psi_i$ and the local error is order d^2 and is therefore only slowly converging.*

$$\underset{\sim}{\Psi}_{i+1} = \underset{\sim}{\Psi}_i + d \cdot \hat{\underset{\sim}{\Psi}},$$

$$\hat{\underset{\sim}{\Psi}} = \frac{1}{6}(\underset{\sim}{K}_1 + 2\underset{\sim}{K}_2 + 2\underset{\sim}{K}_3 + \underset{\sim}{K}_4),$$

$$\underset{\sim}{K}_1 = \underset{\sim}{A}(s_i) \underset{\sim}{\Psi}_i,$$

$$\underset{\sim}{K}_2 = \underset{\sim}{A}(s_i + d/2)(\underset{\sim}{\Psi}_i + \underset{\sim}{K}_1 d/2),$$

$$\underset{\sim}{K}_3 = \underset{\sim}{A}(s_i + d/2)(\underset{\sim}{\Psi}_i + \underset{\sim}{K}_2 d/2),$$

$$\underset{\sim}{K}_4 = \underset{\sim}{A}(s_i + d)(\underset{\sim}{\Psi}_i + \underset{\sim}{K}_3 d).$$

With this scheme the local error is of order d^5. When $\underset{\sim}{A}(s)$ is only defined at discrete points the method of Adams or other multistep methods may be preferable, see Szidarovszky and Yakowitz (1978).

The actual computation uses shooting, the shooting parameter being the frequency σ and the penalty function being the determinant (6.16).

b) New types of topographic waves

All exact models permitted explicit determination of eigenfrequencies and solutions when two mode numbers were given. There exists, however, no simple rule to predict mode numbers for a given frequency interval and thus to discover all eigenmodes in this interval. Furthermore, by increasing the order N of the expansion (4.3) more transverse variability is introduced and, if possible, additional eigenfrequencies and modes are added to the spectrum.

We investigate the spectrum of topographic waves in a second and a third order model. The basin is rectangular with an aspect ratio $r = 0.5$, a parabolic cross section ($q = 2.0$) and a thalweg varying as a $(\sin)^2$. Figures 6.6 and 6.7 display a selection of modes from the spectrum of a second and a third order model, respectively. It is apparent that in the period interval from 35 h to 140 h (corresponding to 45° latitude) a large variety of qualitatively different eigenmodes can be detected. According to the complexity of their modal structure we distinguish three types of eigenmodes.

Type 1 is the well known modal pattern described by all exact models of topographic waves in enclosed basins. It is akin to Ball's solutions (Ball, 1965) and therefore called Ball-type. Both, the linear ($\sigma = 0.155$) and the quadratic ($\sigma = 0.213$) Ball-mode occur in the spectrum and additional eigenmodes are identified as type 1. All exact models for which solutions have been constructed so far, have shown qualitatively similar solutions. Generally, type 1 modes consist of a few large-scale

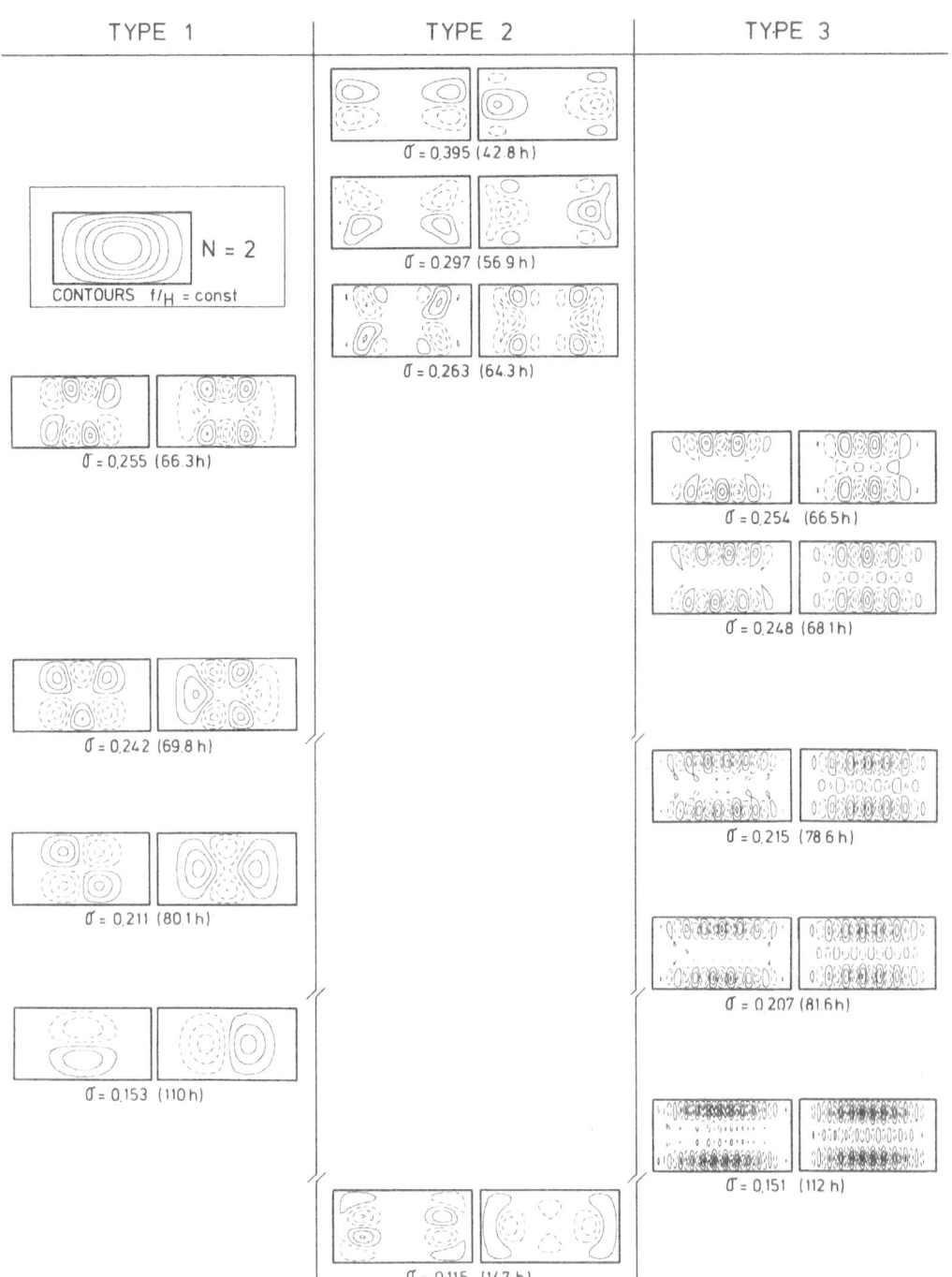

Figure 6.6

Selection through the spectrum containing eigenmodes of a second order model. The contour lines of ψ are plotted for time t = 0 (left) and
t = ¼ T (right). Three types of solutions can be distinguished and cuts
of the vertical lines indicate further modes not shown here. The parameters are: N = 2, r = 0.5, q = 2.0, ε = 0.05, η = 0.01.

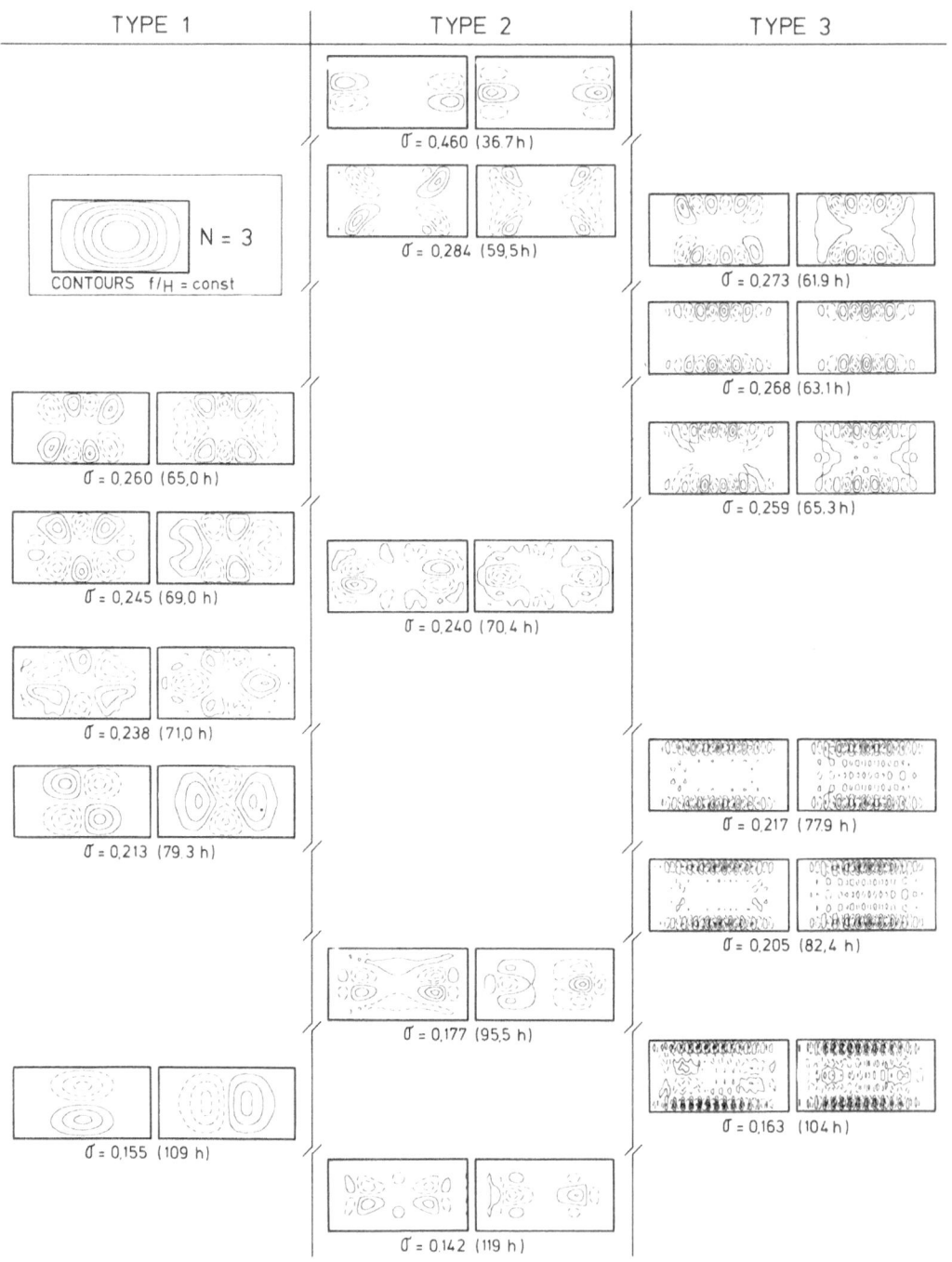

Figure 6.7

Same as Figure 6.6 for a third order model N = 3.

vortices moving counterclockwise around the basin, and the water in the whole basin underlies this wave motion. The rectangular basin, however, appears to sustain also two new types, which so far were unnoticed in other models.

Type 2, with only a few candidates in this frequency interval, can be called *bay-type*. Wave motion is mostly trapped to the long ends of the lake; very weak activity is experienced in the lake center and along its long sides. The pattern shows one or more mid-scale gyres which do not propagate along the entire isobaths (lines of constant f/H) but are rather trapped in the bays. This type arises above the cut-off frequency σ_0 of any mode unit, see *Figure 6.8*, and thus embraces contributions with complex wavenumbers. The amplitudes of these modes are exponentially evanescent in space which makes it understood why bay-type solutions do exist for enclosed basins. The fact that there are eigenmodes with frequencies $\sigma > \sigma_0$ is a new result. These modes were neither detected by the analytic models nor by the crude lake model presented in the previous section[*].

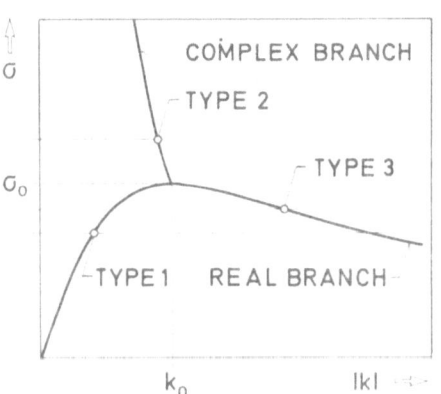

Figure 6.8

Schematic plot of one mode unit of the dispersion relation of topographic waves in a channel with parabolic transverse depth profile. A Nth order model consists of N mode units.

Type 3, eventually, appears most frequently in the spectrum. In contrast to type 2, all wave activity is now trapped along the long boundaries of the basin and consists of a large number of small-scale vortices. The pattern is very similar to that found in straight infinite channels; type 3 is thus named *channel-type*. Along the long sides two seemingly non-interacting beat-patterns are observed. They originate from reflection of wave energy in the bays and a corresponding wavenumber shift, cf. the following chapter.

The modal structure of the different types can be explained with the

[*] *Their determination is very difficult even with high-accuracy integrators. In order to obtain patterns with $|\psi(s,n)| = |\psi(1-s,n)|$ the eigenfrequency need be known up to a relative error of 10^{-7}.*

help of the Rossby dispersion relation in *Figure 6.8*. Type 1 enjoys the property that increasing σ brings about more complex structure since it consists primarily of modes with wavenumbers $k < k_0$. For $k < k_0$ $\partial\sigma/\partial k > 0$ and so the wavelengths of the contributing modes decrease with growing σ. Type 3, the channel-type, on the other hand, reveals the opposite property: the scale of the wave pattern gradually decreases with decreasing frequency. Type 3 solutions are mostly made up of modes with $k > k_0$. In this range, $\partial\sigma/\partial k < 0$ and consequently the wavelengths decrease with decreasing σ, c.f. *Figures 6.6 and 6.7*.

The fact that topographic waves in a rectangular basin occur as bay-trapped modes casts light on the results of Trösch (1984). These seemed to entirely contradict the applicability of analytic models to real basins as anticipated in *Figure 6.1*. Each mode is trapped to one of the bays and does not seem to influence the rest of the basin. The few trapped vortices exhibit roughly the scale of the bay. The rectangular basin, yet a much simpler configuration than Lake of Lugano, reveals equally bay-type modes *together* with the known Ball-type solutions which in the interested period range were not found by Trösch (1984). This model, therefore, links these two different approaches and demonstrates that the propagation of topographic waves in enclosed basins cannot merely be described by those analytically determined modes of exact models that were so far constructed. It is in principle possible and remains to be proved or disproved that type-2 modes also exist in ellipses with parabolic or exponential bottom profiles and that these modes have a period of the same order of magnitude as those above. It is seen that many further questions need to be answered to fully understand the behavior of TW's in enclosed basins. Two facts have, however, transpired: Firstly, the smootheness of the isobaths is essential in enabling global TW-features and, secondly, careful numerical solution procedures are needed to find bay-type modes.

c) *Convergence and parameter dependence*

The quality of approximation strongly depends on the type of wave considered. Ball-type modes have large-scale vortices, and a good representation of these modes with comparatively few basis functions is expected. High orders of expansions are therefore not needed and fast convergence is observed. By contrast channel-type solutions consist of small-scale modes with large wavenumbers. As was shown in section 5.3 convergence is slow for large wavenumbers and this must equally be expected for type 3 modes.

Table 6.3 collects results of a convergence test for the same configuration as in Figures 6.6 and 6.7. Type 1 shows convergence for both, eigenfrequency and stream function; similar but considerably slower convergence is found for types 2 and 3. For type 2 it is particularly difficult to determine the correct distribution of the stream function along the axis, as small changes in the eigenvalue σ result in relatively large changes of the eigenfunction. Thus, high resolution and small step sizes in the numerical integration procedure are needed. Since for ODE high accuracy integrators exist, the channel method allows for some compromise, and this at least explains the superiority of the method in comparison to some other numerical procedures.

Table 6.4 collects the dependence of σ on the aspect ratio and transverse topography for the solutions that correspond to Ball's quadratic mode.

As expected from the behavior of the dispersion relation in a straight infinite channel the slope of the transverse topography has a dominant influence on the values of the eigenfrequency. Steeper profiles (q = 5.0) lower the eigenfrequencies. An equal but weaker effect on Ball-type modes is experienced when the aspect ratio is decreasing. Table 6.4 demonstrates that these modes are much more governed by the transverse depth profile than by the aspect ratio. All this is in line with results obtained from the crude lake model.

Tables 6.5 and 6.6 investigate the influence of the two bathymetric parameters q and r on the three types of basin solutions. Again the topography effect is seen to be more influential. By going from a triangular depth profile (q = 1.0) to a very steep U-shaped profile

Type	N = 1	N = 2	N = 3
1, Ball-type	0.143	0.153*	0.155*
	0.181	0.211*	0.213*
	0.195	0.255*	0.260*
2, bay-type	-	0.297*	0.314
	-	0.263*	0.284*
	-	0.115*	0.240*
3, channel-type	0.151	0.254*	0.273*
	0.142	0.248*	0.268*
	0.111	0.215*	0.253*

Table 6.3
Convergence properties of the eigenfrequencies in a 2:1 basin with q = 2.0, ε = 0.05, η = 0.01. Stars indicate plotted modes in Figures 6.6 and 6.7.

Ball quadratic	r = 0.5	r = 0.4	r = 0.3
q = 1.0	0.267	0.250	0.219
q = 2.0	0.211	0.195	0.170
q = 5.0	0.140	0.123	?

Table 6.4
Topography q and aspect ratio r influencing the eigenfrequency of the quadratic Ball-mode. The parameters are N = 2, ε = 0.05, η = 0.01.

r = 0.5	Ball-type	bay-type	channel-type
q = 1.0	0.200	0.299	0.250
q = 2.0	0.153	0.263	0.232
q = 5.0	0.097	0.175	0.153

Table 6.5

Topography effect on the frequency of the three wave types. The parameters are as in *Table 6.4*.

q = 2.0	Ball-type	bay-type	channel-type
r = 0.5	0.153	0.263	0.232
r = 0.4	0.139	0.267	0.251
r = 0.3	0.118	0.269	0.258

Table 6.6

Aspect ratio effect on the three types. The parameters are as in *Table 6.4*.

(q = 5.0) the eigenfrequencies diminish by up to a factor of 2. As far as the topography effect is concerned, the three types react the same way, yet Ball-type modes are more sensitive to an increase of q.

Table 6.6 demonstrates that basins with a smaller aspect ratio sustain Ball-type waves with decreased eigenfrequencies. This decrease is over-proportional as it is enhanced for smaller aspect ratios. By contrast, bay- and channel-type solutions show an opposite behavior. Decreasing the aspect ratio increases the eigenfrequency; this time the response is under proportional and for bay-type solutions the dependence of σ and r is very small.

d) *The bay-type*

The occurrence of bay-trapped modes in enclosed basins was unexpected and raises further questions concerning the properties of solutions of of the eigenvalue problem (2.22).

When the aspect ratio of the basin is decreased the bay vortices of these modes lie farther and farther apart and we wonder whether these isolated gyres become uncoupled. There are two points to be remarked in this context. Firstly, basins with no symmetry seem to sustain decoupled bay modes as in *Figure 6.1b*. Secondly, with our procedure it is very difficult to determine the parity[*] of these solutions with respect to the long axis of the basin. In this regard very fine resolution is needed to obtain reliable solutions. The basic problem is to bring numerical information through the "dead" zone in the center of the domain. This suggests to consider again semi-infinite channels and to ask the question of a possible existence of *bay-trapped modes*. A partial answer is given in the following chapter.

---

[*] $\psi(s,n)$ has positive or negative parity with respect to s when $\psi(s,n) = \psi(1-s,n)$ or $\psi(s,n) = -\psi(1-s,n)$, respectively.

7. Reflection of topographic waves

In this chapter we intend to investigate the behavior of TW's in channels when they are reflected by a vertical wall or a shore zone. To this
end we consider the propagation of TW's in semi-infinite channels. This
configuration is especially of interest as a limit of elongated basins,
for the distribution mechanism of TW-energy within an enclosed basin may
be supposed to consist primarily of a series of subsequent reflections
at the long ends. This will result in a superposition pattern which eventually will be observed as a basin mode. Alternatively, if a semi-infinite channel should permit bay-modes trapped to the channel end then it
is plausible that similar modes in elongated rectangles could exist in-
dependently at either lake end.

Scrutinizing the properties of one individual reflection will, however,
not only shed light on this problem but equally help us to explain most
of the conspicuous features found in the previous chapter. There, we observed that the crude lake model (vertical walls) did not give rise to
fundamental modes. The wave structure was rather small-scale and no global phase rotation was obtained. Can this result be substantiated by
the study of TW-reflection ? Moreover, in section 6.2, by using merely
phenomenological arguments, we distinguished three types of TW's. These
showed different parameter dependencies; conjectures were put forward
for their *physical* explanation. We assert here that the key to these
answers can also be found by extracting the basic mechanism: the reflection of TW's in a semi-infinite channel.

To date, little is known about the solution of the TW-equation (2.22) in
semi-open domains such as gulfs, harbours etc. Most water wave studies
in such domains are concerned with first-class waves and their behavior
under reflection. Taylor (1922) showed that the energy of an incident
Kelvin wave propagating towards a vertical wall is distributed among a
reflected Kelvin wave *and* a whole spectrum of Poincaré waves. Provided
$\sigma < \sigma_p$ where σ_p is the cut-off frequency of Poincaré waves the latter exhibit a spatially exponential decay and therefore are only important in
the neighbourhood of the reflecting wall. Brown (1973) also constructed
solutions for $\sigma > \sigma_p$ using the method of collocation. For this case, some
of the Poincaré modes are oscillatory and no longer evanescent; they
may also destroy the symmetry of the reflection pattern. A summary is
given in LeBlond & Mysak (1980). Webb & Pond (1986) investigated the
transmission and reflection of a Kelvin wave propagating in a channel

when hitting a bend.

Another important effect which occurs in open domains is called *harbour resonance*. Although to our knowledge there exist no studies of this effect for TW's, considerable knowledge has been acquired for first class waves, see the review article by Miles (1974), Miles & Lee (1975) and Buchwald & Williams (1975).

7.1   Reflection at a vertical wall

A property of our channel technique is the fact that it furnishes solutions with complex wavenumbers in a natural way. This suggests that solutions of the form (5.9) can be found which represent the situation of reflected topographic Rossby waves in a channel. The idea is to superpose several waves with the same frequency; one incident and some reflected waves. The incident and at least one reflected wave have real k[*), and the remaining modes have $Im(k) > 0$; they are important only in a boundary zone where the reflection is induced. The superposition satisfies the boundary condition $\psi = 0$ (no flux) at the reflecting wall.

Consider a semi-infinite channel $s \geq 0$ with a wall at $s = 0$. One particular wave mode forms the incident wave; possible candidates are indicated in *Figure 7.1*. These modes have their group velocity directed towards the wall, and the transverse structure of the incident wave depends on the mode unit to which they belong. Reflected modes which take part in the superposition must not be among the indicated modes and must satisfy the inequality

$$Im\, k \geq 0. \qquad (7.1)$$

With this the superposition and determination of the compound solution is unique and consists of one incident mode and $2N$ reflected modes. This argument relies on the fact that the real branch of the dispersion rela-

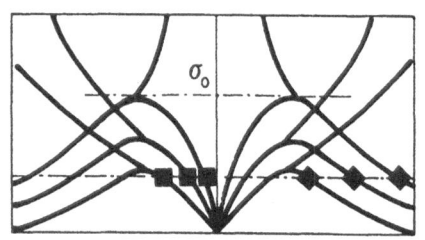

Figure 7.1

Dispersion relation $\sigma(|k|)$ of a third order model in an infinite channel. Possible incident modes with group velocity into the negative s-direction are indicated by

■ for $c_{ph} \uparrow\uparrow c_{gr}$ and ◆ for $c_{ph} \downarrow\uparrow c_{gr}$. Above the cut-off frequency σ_0 all wavenumbers are complex.

---

*) *This implies that $\sigma < \sigma_0$, σ_0 is the cut-off frequency indicated in Figure 7.1, above it all wavenumbers are complex.*

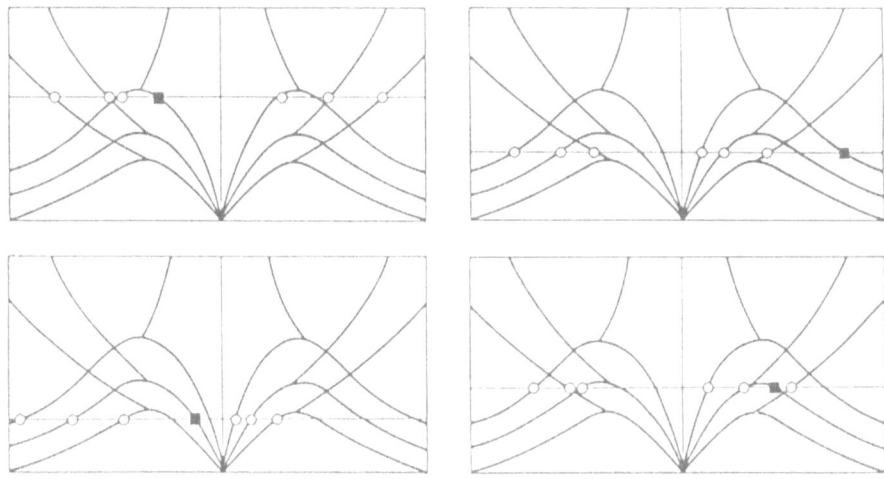

Figure 7.2   Selection of possible incident (■) and reflected (o)
modes in a semi-infinite channel.

tion has domains of k > 0 where ∂σ/∂k < 0 and ∂σ/∂k > 0. It also makes
use of the existence of a complex branch of the mode units. A series of
examples of this selection is shown in *Figure 7.2*. Incident waves are
marked with full squares, reflected modes are shown with open circles.
Those on the real branch have the energy propagating away from the barrier s = 0, those on the complex branch arise in pairs, but actually
represent four complex wavenumbers of which only two have Im k > 0, see
Figure 5.3.

Dropping the harmonic time dependence a general wave in a straight, infinite channel reads

$$\psi = \sum_{\gamma} d_\gamma \ e^{ik_\gamma s} \sum_{\alpha} c_{\alpha\gamma} \ P_\alpha, \tag{7.2}$$

where we have neglected to explicitly distinguish between P_α^+ and P_α^- with
$\alpha = 1,\ldots,N$ and use for simplicity only P_α with $\alpha = 1,\ldots,2N$. A solution, representing wave reflection, is then given by

$$\psi = \psi_{in} + \psi_{out} = \sum_{\alpha=1}^{2N} \psi_\alpha = e^{ik_i s} \sum_{\alpha=1}^{2N} c_{\alpha i} \ P_\alpha + \sum_{\gamma=1}^{2N} d_\gamma \ e^{ik_\gamma s} \sum_{\alpha=1}^{2N} c_{\alpha\gamma} \ P_\alpha, \tag{7.3}$$

with the unknown vector d_γ. The coefficients $c_{\alpha i}$ are known if the frequency σ and corresponding wavenumber k_i of the incident wave are prescribed. They are computed with the methods of chapter 5. Analogously,
to each of the wavenumbers $k_\gamma(\sigma)$ of the reflected waves the correspond-
ing $c_{\alpha\gamma}$ can be computed. Hence k_i, k_γ, $c_{\alpha i}$ and $c_{\alpha\gamma}$ ($\alpha,\gamma = 1,2,\ldots,2N$) are
known.

Imposing the no-flux condition $\psi_\alpha = 0$ at $s = 0$ yields the linear system

$$\sum_{\gamma=1}^{2N} c_{\alpha\gamma}\, d_\gamma = -c_{\alpha i}, \qquad \alpha = 1,\ldots,2N, \tag{7.4}$$

d_γ and $c_{\alpha i}$ are vectors of length $2N$ and $c_{\alpha\gamma}$ is a $(2N \times 2N)$-matrix. Due to the orthogonality of the set $\{P_\alpha\}$ and the modes belonging to different wavenumbers k_γ the matrix $c_{\alpha\gamma}$ is regular and (7.4) can be inverted.

Figure 7.3a displays the wave pattern which results, when a wave belonging to the first mode unit with both phase and group velocities directed towards the wall is reflected. Alternatively, incident phase and group velocities may have different directions as in the second-mode response of *Figure 7.3b*. We have found that the largest portion of the reflected energy lies in the mode with the corresponding wavenumber belonging to the same branch of the dispersion relation (indicated by arrows in the insets of *Figure 7.3*).

Therefore, the mode with the negative of the incident wavenumber is hardly excited, and reflection causes primarily a shift of wavenumber rather than a change of its sign. As a consequence, wave activity remains at the side of the incident wave. What results is a beat pattern with its first "calm" area at approximately $2\pi B / |k_{in} - k_{out}|$ away from the wall. The structure depends on the two main wavenumbers k_{in} and k_{out}. If these differ markedly from each other rather local and small-scale patterns emerge.

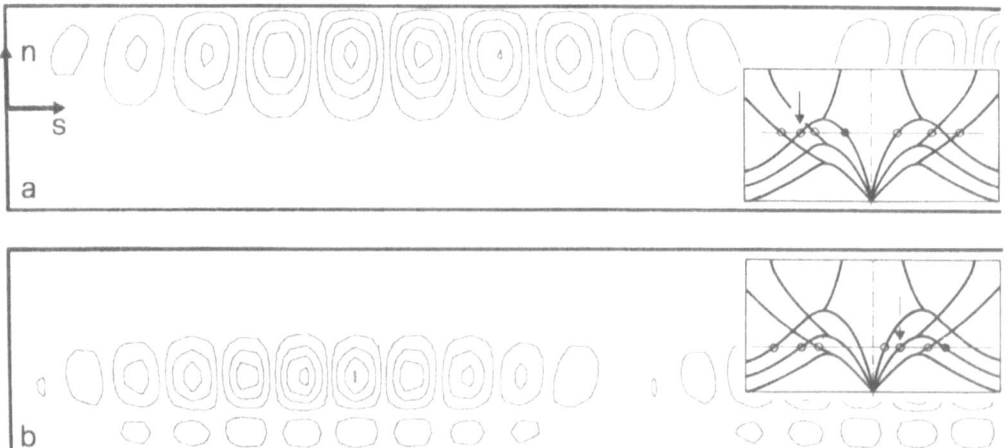

Figure 7.3
ψ-contour lines of a reflection of topographic waves at a vertical wall. The insets explain the composition of the reflection pattern with ●, incident wave and o, reflected wave. The selected parameters are $N = 3$, $\varepsilon = 0.05$, for a) $q = 1.0$, $\sigma = 0.305$ and b) $q = 0.5$, $\sigma = 0.202$.

These results give a better understanding of the basin modes obtained when studying a crude lake model. Comparing *Figure 7.3a and 6.3a* clearly indicates that the basin mode is merely the superposition of two nearly independent reflection patterns which are induced by the two vertical walls. Due to the fact that the discontinuous depth lines prevent wave energy from changing the side in the channel, there are no simple reflection patterns to be expected that occupy the whole channel. Hence, in a semi-infinite channel or even a lake basin the along-axis depth profile at the very end is of crucial importance for the structure of the reflection pattern.

7.2 Reflection at an exponential shore

We now consider a case which has continuous (though not everywhere differentiable depth lines at the end of the channel. In order to keep the convenient algebraic procedure (7.3), (7.4) we let the channel be composed of two sections. Close to the end-wall, for $0 < s < s_0$, the depth increases exponentially as $h(s) = \varepsilon (1 + \frac{1}{\varepsilon})^{s/s_0}$, for $s > s_0$ it is constant. The isobaths no longer intersect the wall but are $\subset$-shaped. Safe a time-dependent factor $e^{-i\omega t}$ the solution then takes the form

$$\psi = \begin{cases} \psi^0 = \sum_{\Gamma=1}^{4N} d_\Gamma \, e^{ik_\Gamma s} \sum_{\alpha=1}^{2N} c_{\alpha\Gamma}^{ex} \, P_\alpha, & 0 < s < s_0, \\[4mm] \psi^\infty = e^{ik_i s} \sum_{\alpha=1}^{2N} c_{\alpha i} \, P_\alpha + \sum_{\gamma=1}^{2N} d_\gamma \, e^{ik_\gamma s} \sum_{\alpha=1}^{2N} c_{\alpha\gamma} \, P_\alpha, & s_0 < s, \end{cases} \qquad (7.5)$$

where k_i is the incident wavenumber, $\{k_\Gamma\}_1^{4N}$ is the whole set of wavenumbers and $\{k_\gamma\}_1^{2N}$ is the restricted set with $\text{Im} \, k \geq 0$ and the group velocity directed away from the end wall, all corresponding to σ. Superscripts 0 and ∞ denote the domains $0 < s < s_0$ and $s_0 < s$, respectively. $c_{\alpha\Gamma}^{ex}$ is the $(2N \times 4N)$-matrix corresponding to (5.8) but for the case $h'/h = \text{const} \neq 0$. The stream function ψ must be continuous and differentiable at $s = s_0$ and vanish at $s = 0$. Thus, for $\alpha = 1, \ldots, 2N$[*)]

$$\left. \psi_\alpha^0 \right|_{s=0} = 0 \qquad : \quad c_{\alpha\Gamma}^{ex} \, d_\Gamma = 0,$$

$$\left. \psi_\alpha^0 \right|_{s=s_0} = \left. \psi_\alpha^\infty \right|_{s=s_0} \quad : \quad e^{ik_\Gamma s_0} c_{\alpha\Gamma}^{ex} \, d_\Gamma = c_{\alpha\gamma} \, e^{ik_\gamma s_0} \, d_\gamma + c_{\alpha i} \, e^{ik_i s_0}, \qquad (7.6)$$

$$\left. \frac{\partial}{\partial s} \, \psi_\alpha^0 \right|_{s=s_0} = \left. \frac{\partial}{\partial s} \, \psi_\alpha^\infty \right|_{s=s_0} \quad : \quad ik_\Gamma \, e^{ik_\Gamma s_0} c_{\alpha\Gamma}^{ex} \, d_\Gamma = ik_\gamma \, e^{ik_\gamma s_0} c_{\alpha\gamma} \, d_\gamma + ik_i \, c_{\alpha i} \, e^{ik_i s_0},$$

---

*) We now omit the summation signs over $\Gamma = 1, 2, \ldots, 4N$ and $\gamma = 1, 2, \ldots, 2N$.

with the 6N unknowns d_Γ and d_γ must be satisfied. This can be written as

$$
\begin{bmatrix}
c_{\alpha\Gamma}^{ex} & 0 \\
c_{\alpha\Gamma}^{ex} e^{ik_\Gamma s_0} & -c_{\alpha\gamma} e^{ik_\gamma s_0} \\
ik_\Gamma c_{\alpha\Gamma}^{ex} e^{ik_\Gamma s_0} & -ik_\gamma c_{\alpha\gamma} e^{ik_\gamma s_0}
\end{bmatrix}
\begin{bmatrix}
d_\Gamma \\
d_\gamma
\end{bmatrix}
=
\begin{bmatrix}
0 \\
c_{\alpha i} e^{ik_i s_0} \\
ik_i c_{\alpha i} e^{ik_i s_0}
\end{bmatrix},
\tag{7.7}
$$

and the vectors d_Γ and d_γ are determined by inverting (7.7).

Figure 7.4 shows solutions ψ for a composed channel; two significant differences to *Figure 7.3* are observed. Now, there is wave activity also in the opposite half of the channel corresponding to the negative of the incident wavenumber. This amounts to a weak *leakage of wave energy* by reflection into the other channel domain (*Figure 7.4a*). However, probably owing to the non-smoothness of the isobaths at s_0 it is comparatively weak and most of the reflected wave activity remains on the incident side.

Figure 7.4b shows a reflection pattern of lower frequency, k_{in} and k_{out} lie farther apart and therefore more local and complicated structures result. Moreover, at the beginning of the reflecting shelf $(s \approx s_0)$ *wave intensification* is observed. These specific results demonstrate that the global wave pattern is very sensitive to the basin shape and the depth profile

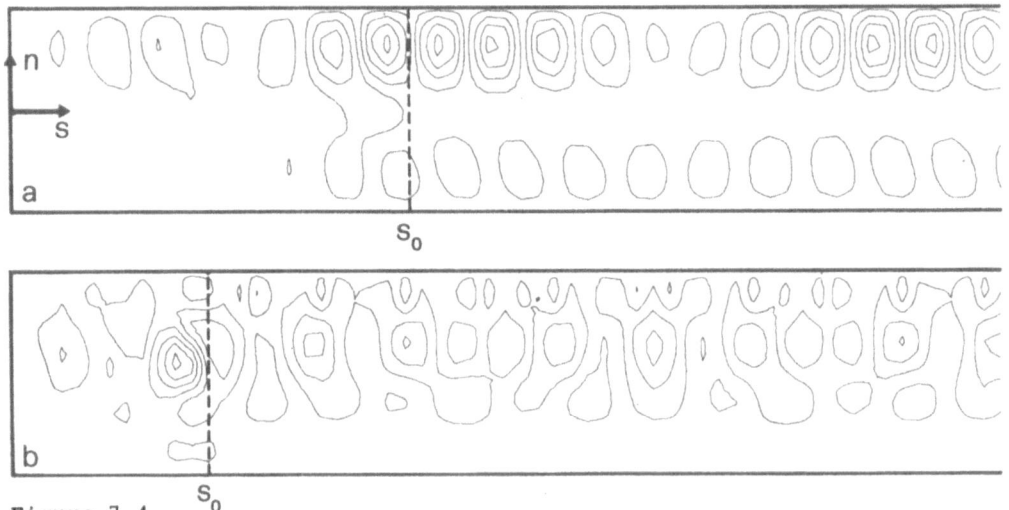

Figure 7.4

Reflection pattern in a composed channel. For $0 < s < s_0$ the depth varies exponentially along the axis whereas it is constant for $s > s_0$. This connects the isobaths of both channel domains $n > 0$ and $n < 0$ and enables wave energy to leak into the opposite domain in the course of reflection. The selected parameters are $\varepsilon = 0.05$, for a) $N = 2$, $q = 2.0$, $\sigma = 0.260$, $s_0 = 2.0$ and b) $N = 3$, $q = 0.5$, $\sigma = 0.200$, $s_0 = 1.0$.

at the channel end. However, these results still do not explain the distinction of TW's into three different basin types as suggested in section 6.2. We would like to have these explained e.g. as special cases of three different reflection patterns.

7.3 Reflection at a sin²-shore

This section closely follows the analysis in section 6.2. The procedure is, however, slightly more complicated since we must construct solutions in an *open* domain.

a) *Numerical method*

The domain of interest is a semi-infinite channel with the depth profile (6.2) and a thalweg depth

$$h(s) = \begin{cases} \eta + \sin^2 \dfrac{\pi s}{2s_0}, & 0 < s < s_0, \\ 1 + \eta, & s_0 < s. \end{cases} \tag{7.8}$$

This profile is smooth at s_0 and the slope parameter $S(s) \equiv h^{-1} \, dh/ds$ takes the form

$$S(s) = \begin{cases} \dfrac{\pi \sin(\pi s/s_0)}{2 s_0 (\eta + \sin^2(\pi s/2s_0))}, & 0 < s < s_0, \\ 0, & s_0 < s. \end{cases} \tag{7.9}$$

The solution ψ in the two domains is given by

$$\underset{\sim}{\psi} = \begin{cases} \underset{\sim}{\psi}^0 = \exp\left(\int_0^s \underset{\sim}{A}(\hat{s}) \, d\hat{s} \right) \underset{\sim}{\psi}^0(0) = \underset{\sim}{E}(s) \, \underset{\sim}{\psi}^0(0), & 0 < s < s_0, \\ \underset{\sim}{\psi}^\infty = \underset{\sim}{\psi}_i^\infty + \displaystyle\sum_{\gamma=1}^{2N} \underset{\sim}{\psi}_\gamma^\infty \, D_\gamma, & s_0 < s, \end{cases} \tag{7.10}$$

where (6.5) and (6.12) have been used. $\underset{\sim}{\psi}_i^\infty$ is the incident mode with wave number k_i and, in real notation, has the form

$$\underset{\sim}{\psi}_i^\infty = (\text{Re}\,\psi_i, \ \text{Re}\,\dot{\psi}_i, \ \text{Im}\,\psi_i, \ \text{Im}\,\dot{\psi}_i), \tag{7.11}$$

which is a vector with 8N components; one such component, e.g. $(\text{Re}\,\psi_i)_\alpha$ is given by Re $\exp(i\,k_i\,s)\,c_{\alpha i}\,P_\alpha$. If D_γ and ψ_γ^∞ are also separated into real imaginary parts, they have the form[*]

---

[*] *The extended formulations (7.12) do not contain more information than the form (6.5) and only account for the characteristics of the complex multiplication. Capital subscripts R and I denote real and imaginary parts, respectively.*

$$\underset{\sim}{\psi}_\gamma^\infty = \left[\begin{array}{c|c} \text{Re } \psi_\gamma & -\text{Im } \psi_\gamma \\ \text{Re } \dot\psi_\gamma & -\text{Im } \dot\psi_\gamma \\ \hline \text{Im } \psi_\gamma & \text{Re } \psi_\gamma \\ \text{Im } \dot\psi_\gamma & \text{Re } \dot\psi_\gamma \end{array}\right] , \qquad D_\gamma = \left[\begin{array}{c} D_{R\gamma} \\ D_{I\gamma} \end{array}\right] . \tag{7.12}$$

As was the case for the incident wave, ψ_γ consists of 2N components, each of which has the form

$$(\underset{\sim}{\psi}_\gamma)_\alpha \equiv e^{ik_\gamma s} \, c_{\alpha\gamma} \, P_\alpha, \qquad (\underset{\sim}{\dot\psi}_\gamma)_\alpha \equiv ik_\gamma \, e^{ik_\gamma s} \, c_{\alpha\gamma} \, P_\alpha, \qquad \alpha = 1,2,\ldots,2N$$

and wavenumbers are restricted such that $\text{Im } k_\gamma \geq 0$.

The representation (7.10) has 8N real unknowns, $\psi^0(0) = (0, \dot\psi_R^0(0); \; 0, \dot\psi_I^0(0))$ and D_γ. These are determined with the help of the matching condition at $s = s_0$, viz.

$$\left.\underset{\sim}{\psi}^0\right|_{s=s_0} = \left.\underset{\sim}{\psi}^\infty\right|_{s=s_0} \quad : \quad E(s_0) \, \underset{\sim}{\psi}^0(0) = \underset{\sim}{\psi}_i^\infty + \sum_\gamma \underset{\sim}{\psi}_\gamma^\infty \, D_\gamma ,$$

or more precisely,

$$\left[\begin{array}{c|c} \underset{\sim}{E}_{12} & \underset{\sim}{E}_{14} \\ \cdot & \cdot \\ \cdot & \cdot \\ \underset{\sim}{E}_{42} & \underset{\sim}{E}_{44} \end{array}\right| \left. -\underset{\sim}{\psi}_\gamma^\infty\right] \left[\begin{array}{c} \dot\psi_R^0(0) \\ \dot\psi_I^0(0) \\ D_\gamma \end{array}\right] = \left[\begin{array}{c} \underset{\sim}{\psi}_i^\infty \end{array}\right] , \tag{7.13}$$

and the calculation of the $\underset{\sim}{E}_{ij}$'s is described in the text below (6.15). The computational scheme therefore requires, firstly, numerical integration by a Runge-Kutta method to obtain the $\underset{\sim}{E}_{ij}$'s and secondly, an algebraic procedure to calculate both $\underset{\sim}{\psi}_\gamma^\infty$ and, for a preselected incident wavenumber k_i, the corresponding $\underset{\sim}{\psi}_i^\infty$.

b) _Reflection patterns_

We learn from (7.10) that 2N+1 modes are superposed which make up the solution $\underset{\sim}{\psi}^\infty$ far away from the reflecting zone. It is of particular interest to determine the reflection coefficients R_γ corresponding to the individual modes with wavenumber k_γ. Usually, these are calculated with the help of an energy argument: R_γ then is proportional to the averaged total energy contained in the mode k_γ. As section 5.4 has revealed, any attempt to draw conclusions concerning the energy content of TW-motion is ambiguous when considerations are restricted to a barotropic formulation. This is so, because the averaged velocity field does not account for the energy content due to vertical velocity variations and therefore is always a _lower bound_. Hence, we propose another procedure.

The measure of "strength" of the contributing modes is selected by scaling the maximum value of the modulus of the stream function ψ_γ with the maximum value of that of the incident mode ψ_i. More precisely, we define R_γ as

$$R_\gamma \equiv \frac{\displaystyle\max_{n \in [B^+, B^-]} \left| (D_{R\gamma} + i D_{I\gamma}) \sum_{\alpha=1}^{2N} c_{\alpha\gamma} P_\alpha(n) \right|}{\displaystyle\max_n \left| \sum_{\alpha=1}^{2N} c_{\alpha i} P_\alpha(n) \right|} . \tag{7.14}$$

We have calculated the reflection coefficients R_γ for a second and a third order model. The former has already revealed remarkable results which are demonstrated in *Figure 7.5*. It shows R_γ of the two possible[*] reflected modes as functions of the frequency. The reflected modes are induced by the incident mode ■ which has $c_{gr} \uparrow\uparrow c_{ph}$ towards $s = 0$.

When solving (7.13) two cases have to be considered. If $\sigma > \sigma_0$ there exist no modes which are periodic in space, i.e. Im $k \neq 0$ for all k. Consequently we cannot define an incident mode as in (7.10). Setting $\psi_i^\infty = 0$, (7.13) allows a non-trivial solution if

$$\det [\underset{\sim}{E}(s_0), - \underset{\sim}{\psi}_\gamma^\infty] = 0. \tag{7.15}$$

On the other hand, when $\sigma < \sigma_0$, $\underset{\sim}{\psi}_i^\infty \neq 0$ and (7.13) is invertible provided the determinant does not vanish.

Calculations have shown that there are indeed real frequencies $\sigma > \sigma_0$ satisfying (7.15). Consequently, there exists a *discrete spectrum for $\sigma > \sigma_0$*. and a *continuous spectrum for $\sigma < \sigma_0$*. The contour lines of the stream function (7.10) for different frequencies are also plotted in *Figure 7.5*. Corresponding to the terminology used in Quantum Mechanics we call the waves which belong to the discrete spectrum *bound states* of TW's in the semi-infinite channel whereas the waves for $\sigma < \sigma_0$ are *free states* of the system. This terminology is very appealing and obviously applies here well as inspection of the stream functions in *Figure 7.5* reveals.

The bound states must be identified with the type 2 waves (bay-modes) found in the improved lake model in section 6.2. Indeed, the frequencies $\sigma = 0.395$ are the same and when ignoring in the rectangle the stream function at the far end $s = 1$ the mode structures are alike, see *Figure 6.6*. We therefore conclude that the occurrence of the bay-mode in the rectangular basins for $\sigma \sim \sigma_0$ is due to two trapped bound states of TW's in

---

[*] *A possible reflected mode has $c_{gr} = \partial\sigma/\partial k$ directed away from $s = 0$, i.e. towards $s = +\infty$ and Im $k = 0$.*

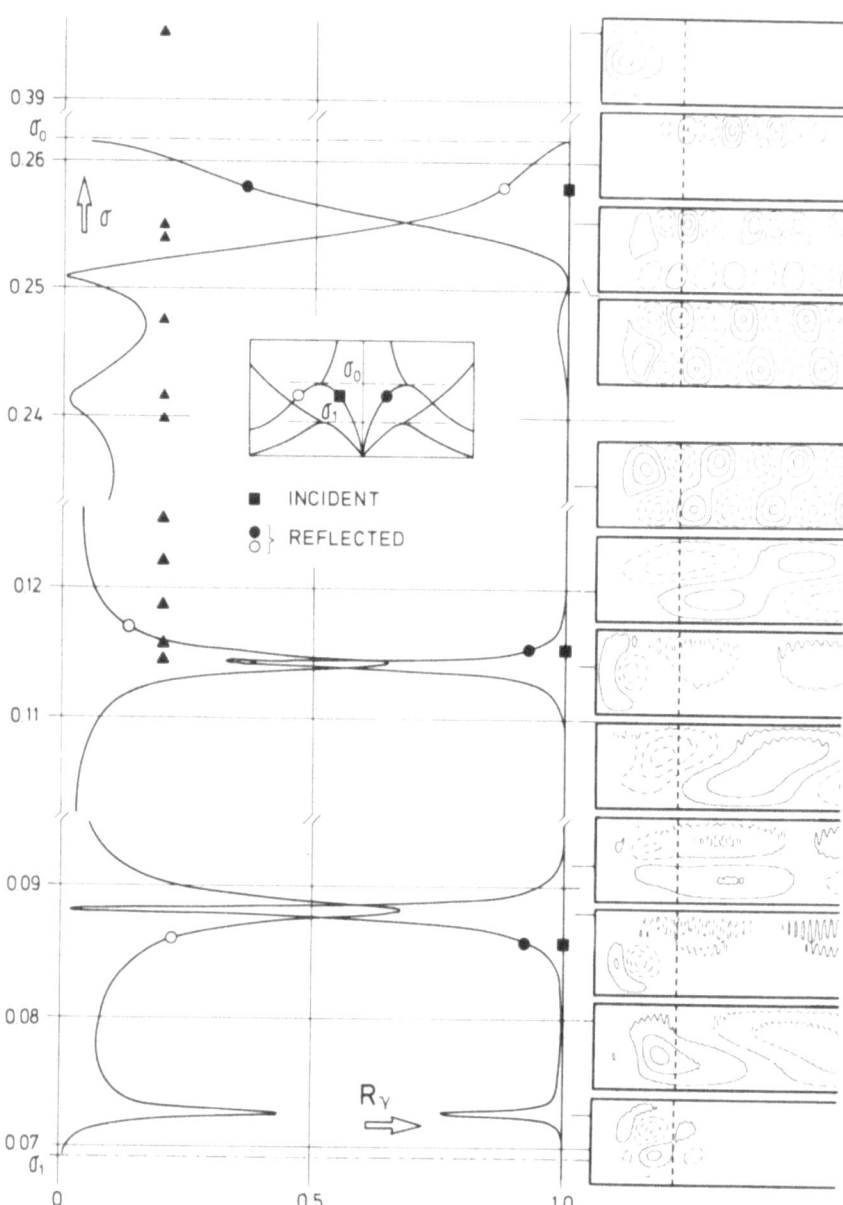

Figure 7.5

Reflection coefficients and stream function patterns in sub-
domains of the frequency interval $[\sigma_1, \sigma_0]$ of the two reflected modes ● and o, respectively. The coefficient of the incident mode ■ is scaled to 1 and both c_{gr} and c_{ph} are directed towards the reflecting wall. ▲ indicate lake solutions
for $\sigma > 0.11$ corresponding to *Figure 6.6*. The inset explains
the position of the modes within the dispersion relation and
the parameters are $N = 2$, $r = 1$, $q = 2$, $\varepsilon = 0.05$, $\eta = 0.01$,
$s_0 = 1$ (dashed line), $M = 50$ for $\sigma > 0.2$ and $M = 200$ for $\sigma \leq 0.2$.

either lake bays at $s = 0$ and $s = 1$. The stream function of this mode consists of 2N modes k_γ with $\text{Im } k_\gamma > 0$ for $s > s_0$ and is spatially evanescent.
The longer a lake basin is, the weaker will be the coupling of the bound
modes in the respective bays. The two additional bay-modes shown in *Figure*
6.6 at $\sigma = 0.297$ and $\sigma = 0.263$ are also originating from bay-trapped topographic waves not shown in *Figure 7.5*.

The fact that equation (2.22) has a discrete spectrum above $\sigma > \sigma_0$ consisting of bound
states resolves the seeming controversy formulated in section 6.1. In particular, in
elongated lakes with very steep transverse topography ($q \leq 10$ for Northern Lake of Lugano
as determined by Bäuerle, 1986) this new result is of importance. Let us estimate the
frequency of the quadratic Ball-mode of the elongated Northern Lake of Lugano. The basin is 17 km long and has an approximate width of 1.5 km. This gives an aspect ratio
of $r = 1.5/17 \approx 0.088$. Using (3.15) and (3.22) yields the estimate

$$\sigma \approx 0.049, \quad T \approx 350 \text{ h}. \tag{7.16}$$

Remember that the topography of the lake has a markedly steeper profile than the parabolic used in the Ball-model. Due to the conspicuous topography effect, (7.16) is certainly an overestimate for σ. Periods would therefore have to be expected to be even
longer. Measurements, however, indicate a distinct signal at around 74 h, clearly far
above the cut-off frequency σ_0 for this basin. One possible new interpretation is thus
put forward, and it seems reasonable that the 74 h-signal could be the trace of a *bay-*
trapped topographic wave of one of the bays at Melide, Lugano or Porlezza, see *Figure*
6.1. Although, the bay-modes have been constructed for a 2:1 basin, these results still
apply for more elongated lakes. For constant s_0 only the topography parameters p and q
determine the frequency of the bay-mode. Decreasing p and increasing q lowers σ considerably. So, a bay-mode with $\sigma = 0.395$, $T = 42.8$ h can easily be brought into accor-
dance with the observed 74 hours. A further argument supporting this interpretation is
the fact that spectral peaks of temperature time series of moorings at the Melide end
(see *Figure 1.6*) have this maximum at periods which are generally slightly larger than
74 h; alternatively, the corresponding peak for the Porlezza mooring is at a slightly
smaller period (compare *Figures 1.6 to 1.8*). The difference could be interpreted as
being due to two independent bound modes that are generated by the different topographies at the two lake ends. The FE-results of Trösch (1984) support this interpretation, see *Figure 6.1*. Mysak et al. (1985), however, also list limited facts which
conflict with this view. Giving a final answer would require data which would uncover
the spatial structure much more clearly.

Starting from σ_0 and decreasing σ we observe that the wave pattern un-
dergoes considerable alterations which correspond to changes in the relative strength of the two reflected modes. More precisely, as R_o decreases $R_\bullet$ increases. For $\sigma < 0.25$ R_o oscillates weakly whilst gradually decreasing and $R_\bullet \gtrsim 0.98$. This can be verified by considering the associated stream functions. For $0.254 \leq \sigma < \sigma_0$ the reflected wave mainly
consists of the o-mode. What evolves is a *beat pattern* at the same chan-
nel-side where the incident mode is located. The increase of $R_\bullet$ manifests itself as a growing *leakage* of wave activity into the opposite
channel side, because the $\bullet$-mode has $k = -k_i$. For $0.120 \leq \sigma \leq 0.254$ $R_\bullet$
is dominant, and this is clearly visible in the wave patterns. The dis-
persion relation has $\partial\sigma/\partial k > 0$ for this reflected mode and consequently,
increasing wavelengths accompany decreasing frequencies. At $\sigma = 0.115$ a

remarkable *resonance* is discovered: Two coinciding peaks give rise to a local minimum and maximum for $R_\bullet$ and R_o, respectively. Looking at the wave pattern suggests that this again is a bay-trapped mode. Contrary to the trapped modes with $\sigma > \sigma_0$ which are true bound states, this mode has also a non-vanishing periodic contribution in $s > s_0$. The pattern is, however, a bay-mode or type-2 wave because the characteristic structure is due to the modes with $\text{Im } k_\gamma > 0$ belonging to the *second* mode unit which still has a complex branch for $\sigma_1 < \sigma < \sigma_0$ (see inset).

The resonance $\sigma = 0.115$ coincides with an eigenfrequency in the closed basin as indicated with ▲. The structure agrees well with that shown in *Figure 6.6*. Below the resonance the component $R_\bullet$ dominates R_o again and large-scale TW's are observed. There is a further resonance at $\sigma = 0.088$.

For $\sigma < \sigma_1$ all modes have $\text{Im } k_\gamma = 0$ in this second order model and no further bay-modes can be expected. Instead of this, contributions of the real branch belonging to the second mode unit are possible. *Figure 7.6* displays the reflection coefficients for the frequency interval $[0.052, \sigma_1]$. All reflection coefficients change smoothly and, as expected, no resonances occur. For $0.063 < \sigma < \sigma_1$ $R_\blacktriangle$ belonging to the second mode unit is dominant (see inset for an explanation of the symbolic subscripts). For lower frequencies the influence of the second mode unit is comparatively weak. Comparing *Figure 7.5* and *Figure 7.6* reveals, that close to the critical frequencies σ_0 and σ_1 energy is distributed among several modes, whereas for other frequencies most of the reflected energy is contained in the ●-mode. This is the mode with the negative of the incident wavenumber.

So far, we have studied the reflections of TW's, when the incident wavemode ■ belongs to the first mode unit and has c_{ph} ↑↑ c_{gr} towards the reflecting zone. We also investigated the situation for an incident mode with c_{ph} ↓↑ c_{gr}. For this case, the graph of *Figure 7.5*, qualitatively looks the same except that the curves R_o and $R_\bullet$ are interchanged. The position of the two conspicuous resonances is unchanged.

Figure 7.7 collects the results of importance. The incident mode with c_{ph} ↓↑ c_{gr} has its wave crests at the opposite side of the channel. Energy is propagating towards s_0 whereas the phase propagates away from it. These two cases distinguish two different types of reflection patterns, type 1 and type 3. Type 1 has a large scale structure with increasing wavelengths for decreasing σ. Conversely, type 3 exhibits a small-scale pattern which is intensified for decreasing frequencies. The distinction

of these types and their individual properties agree with the classifi-
cation suggested in section 6.2. There, we only were able to make the
distinction plausible by phenomenological arguments. We now have discovered a *physical* explanation for the occurrence of bay-modes, Ball-modes
and channel modes in enclosed basins. Comparing *Figure 7.7* with *Figure 6.6*

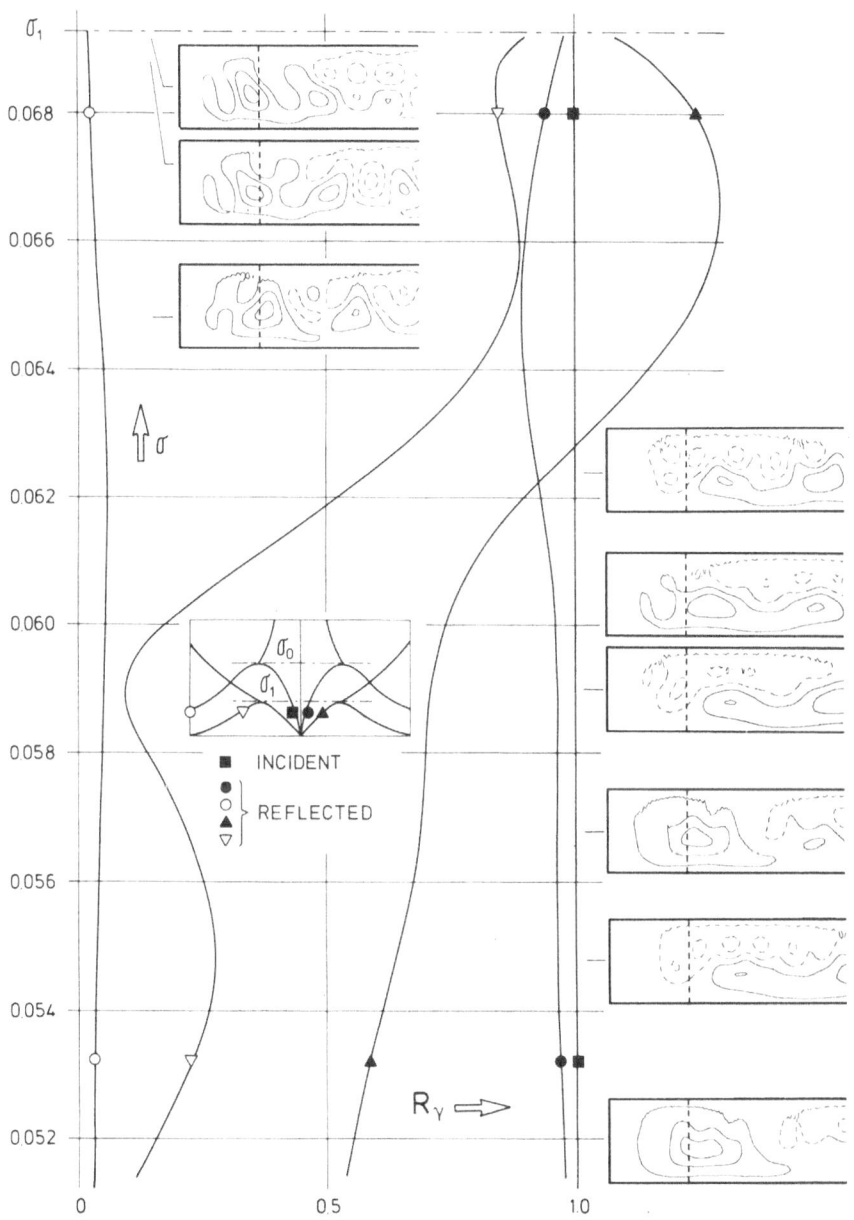

Figure 7.6

Reflection coefficients and stream function patterns for $0.052 \geq \sigma > \sigma_1$
for the four reflected modes. The coefficient $R_\blacksquare$ of the incident mode is
scaled to 1. The parameters are as in *Figure 7.5* and the inset explains
the modes.

makes it clear:

(i) The type 1-modes or *Ball-modes* originate from a sequence of reflections at the lake ends which are induced by an incident wave with $c_{ph} \uparrow\uparrow c_{gr}$. For an appropriately selected frequency, i.e. the eigenfrequency, the pattern is not evanescent in time and a Ball-mode survives.

(ii) The basin solutions classified as type 2 or *bay-modes* are due to the conspicuous resonances observed in *Figure 7.5*. As *Figure 7.7* demonstrates the structure in the bay is only weakly influenced by the incident mode.

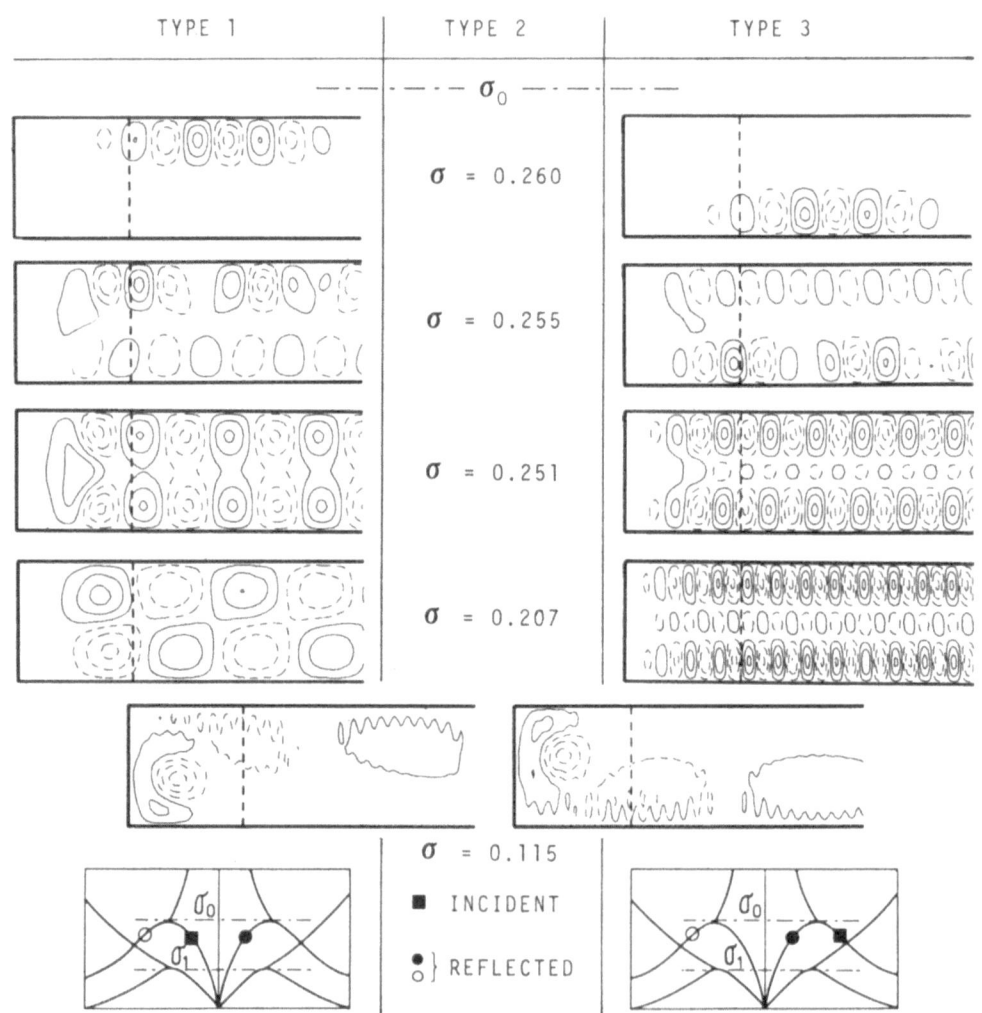

Figure 7.7   Reflection patterns indicated by an incident wave with $c_{ph} \uparrow\uparrow c_{gr}$ (type 1) and $c_{ph} \downarrow\uparrow c_{gr}$ (type 3), respectively. The mode at the resonance $\sigma = 0.115$ constitutes type 2. The parameters are as in *Figure 7.5*.

(iii) Finally, the *channel-modes* or type 3-waves of *Figure 6.4* can be ex-
       plained as the result of a sequence of reflections at the lake
       ends which are induced by a mode with $c_{ph} \downarrow \uparrow c_{gr}$. Contrary to the
       Ball-modes, the spatial scale decreases with decreasing frequency.

These results justify and strengthen the statements which were made in
section 6.2. They provide a more precise and broader understanding of
TW's in channels and lakes. It is now clear that the models to which
(some) exact solutions were presented in chapter 3 do not exhibit the
complete variability of TW's in basins but provide us only with Ball-mode
solutions. These often do not suffice for a reliable interpretation of
field measurements. As the model of Mysak et al. has shown, the ellipse
which could model the 74 h-signal had a far too large aspect ratio. This
discrepancy seems to be removed if the signal is interpreted as a bay-
trapped mode with a frequency that exceeds the cut-off frequency for the
particular basin.

On the other hand, what has been conjectured at the end of chapter 6 is
now made clear in a quantitative manner. The existence of three distinctly different wave types is a natural consequence of the typical dispersion relation of topographic Rossby waves. The conspicuous eigenmodes in
the rectangular basin can be understood in terms of *reflections* of TW's at
either shore-zone. Depending on the structure of the incident wave the
corresponding type is established. All parameter dependencies are expli-
cable with the help of this correspondence.

In elongated lakes the quantities determining the TW-features may, perhaps be listed as follows:

Firstly, the *transverse depth-profile* fixes the frequency range, in which
solutions can be expected. We draw this conclusion from the conspicuous
topography dependence of the frequency illustrated in *Figure 5.8 and Tables
1.1, 3.2, 3.4, 3.6, 5.1, 6.1, 6.2, 6.4, 6.5.* The larger topography gradients
for a fixed maximum depth are, the lower will be the frequencies. There-
fore, σ is strongly influenced by $h^{-1}|\nabla h|$. Secondly, the form of the *lake
ends* is of particular importance as far as the structure of the solution
is concerned. This determines whether a Ball-, bay- or channel-type wave
will occur. Thirdly, it should not be forgotten that TW's are wind-gene-
rated. Depending on the scale of the exciting force the lake basin will
respond differently. Small-scale driving forces will preferably excite
bay-modes or channel-modes whereas large-scale wind forces may produce
Ball-modes.

8. Review and outlook: Restrictions of this study – A list of unsolved problems

8.1 A brief summary

In these notes we hope to have provided the reader with a state of the
art report on the behavior of free topographic Rossby waves in channels
and enclosed basins. As mentioned in the preface, we claim to have
achieved a considerable advancement in the understanding of these waves.
In particular, it was demonstrated that, besides the free travelling
waves, the topographic wave equation admits bound states. These states
consist of wave activity which is essentially restricted to a limited
subdomain near a long end of a basin. For true boundedness of these states wave activity decays exponentially as one moves away from the region of it; incomplete boundedness is characterized by a finite nonzero
far field leakage. These modes require a smooth sloping bottom at a near
shore zone but may otherwise exist in a finite domain or a semi-infinite channel.

The other free travelling wave solutions of the TW-equation are of two
further types. The Ball-type modes consist of a few large scale gyres
which fill the entire basin and rotate counterclockwise (on the Northern hemisphere) around the basin. Most classical solutions that are
known were of this type and, because of the obvious vortex structure of
the wave motion, TW's are also called vortex modes. Finally, when the
basin is filled with a large number of small scale gyres, then we have
the third type of TW's. They are governed mainly by the bathymetric
gradients at the channel side walls. All these wave types have eigenfrequencies in a very narrow frequency band, and a separation of the mode
types according to an ordering of the frequencies has not been possible.

These inferences have been reached with the linearized TW-equation,
which, strictly, applies only in a homogenous water body. Moreover, we
have neglected to discuss many other important aspects. It is thus worthwhile to point out explicitly the limitations of the analysis and to
draw the readers attention to unsolved challenging problems.

8.2 Validity of TW-solutions

Apart from the application of the rigid-lid assumption, which elimina-
tes surface waves, the most obvious omission is that of a barotropic-
baroclinic coupling. The problem was analysed in principle in Chapter 2

(sections 4 and 5) where it was shown that this barotropic-baroclinic coupling was weak: in other words, to first order, the internal Kelvin and Poincaré waves would be influenced by the TW's but not vice versa. For bay-type modes such a coupling may become stronger. When the eigen-frequencies of the bay-modes exceed the cut-off frequency σ_0, their values can enter a range in which processes with a different physical origin are important and dominating. In this case the frequency ranges of the different physical processes overlap and a coupling of the TW's with gravity waves is likely. It would be interesting to know to what extent the total energy is distributed among the TW's and the gravity waves, but also how these waves are modified by the coupling.

A further, perhaps more hidden, limitation is the fact that solutions to our TW-equation may be constructed for conditions which invalidate the applicability of the shallow water equations. We have seen that when $|\nabla h/h|$ or the wavenumber k is large, or when the mode number is high the stream function is either shore trapped or may oscillate rather violently in space: In other words, typical wavelengths are rather small, while associated periods remain large. This indicates that in these instances either the shallowness or the linearity assumption may have to be dropped. Furthermore, because of the large horizontal velocity gradients arising in these cases, viscous effects may also become significant. The orders of magnitude and the relative proportions of these frictional and nonlinearity effects are, of course, still unknown.

8.3 Single or multivalued dispersion relation

All numerically determined dispersion relations in this book are multi-valued, i.e. to every value of σ there belongs a wave with a short wavelength and another one with a long wavelength. In Appendix C this is proved to be a general property of TW's in channels provided that $|\nabla h/h|$ is finite everywhere. A similar proof was also given for shelf waves by Huthnance (1975). On the other hand, numerical solutions (Bäuerle, 1986 and Bäuerle & Hutter, 1986) have indicated that for channels with extremely small shore depths ($|\nabla h/h|$ large) this property of the dispersion relation could no longer be detected. The example of a single-step shelf in Chapter 3 demonstrated monotonicity of the dispersion relation. Our approximate integration technique using the shape function expansion with the trigonometric functions does not allow us to handle such cases. However, several puzzling questions ought to be answered in this regard, as e.g.: Are the dispersion relations always single valued when $|\nabla h/h|$ is unbounded within the domain of solution of the equation ? Does a cut-off frequency exist ? Can our solution technique using shape function expansions be adopted in this singular case ?

Connected with these questions are further intriguing, partly unanswered questions. Are reflections of TW's possible in configurations with a monotone dispersion relation ? How do such reflection patterns look like ? In this context Bäuerle (1986) wonders whether eigenmodes in enclosed domains are possible at all. The answer is: yes they are. We have constructed an analytical solution for an elliptical basin whose depth is constant within an interior confocal ellipse and again within the outer elliptical ring but suffers a jump across the ellipse separating the two domains. This configuration can sustain free TW's.

Further, we may wonder whether bay-type modes will also esist when dispersion relations are monotone. Although there exists a complex branch of the dispersion relation above the cut-off frequency, we do not know at the moment whether such configurations can sustain bay-modes.

8.4 Bay-type modes

These modes came as a surprise, but *a posteriori* their occurrence is rather obvious. We explained them as resonance phenomena of the particular bay or enclosed basin. Further studies should investigate the relevant bathymetric parameters and how they influence frequency and structure of these conspicuous bay-modes. In order to substantiate their existence an analytic model exhibiting bay-modes is desirable. We think of a constant-depth, semi-infinite channel with a shore-zone which could sustain possible bay-modes. On the other hand, from an asymptotic argument their existence can be foreseen. If in a semi-infinite channel the slope parameter $|\nabla h/h|$ is finite close to the long ends but asymptotically small elsewhere, then in the outer region ψ is governed by the Laplacian. In the shore-near zone the full TW-equation must be solved and this interior solution must be matched to the outer solution. If the matching conditions can be satisfied for a non-trivial solution a true bay-mode is found. This is so because the solution of the Laplacian is asymptotically evanescent as $x \to \infty$ and there is no energy leakage. Energy leakage can only occur when the far-field is able to support wave motion i.e. when the slope parameter in the far-field can not be neglected.

8.5 A list of unsolved problems

The above discussion suggests a variety of problems that remain to be solved; there are many more, perhaps less fundamental but nevertheless of practical importance. Here is a list of such problems:

a) *On the computational side*

(i) Topographic wave modes in rectangular basins were only determined for a limited number of topographic parameters q, ε, η, p (see Chapter 6). It would in particular be important to investigate the quantitative influence of variations of these parameters on the mode structure and eigenfrequencies.

(ii) Even though our numerical solution procedure was motivated and developped for *curved* elongated basins all applications in these notes were restricted to straight Cartesian coordinate axes. Analyses of curvature effects were demonstrated only with confocal elliptical coordinates for which the Cartesian coordinate correspondence principle was applicable. The quantitative dependence of TW's on curvature effects under more general situations is still not known.

(iii) Our analysis was performed for models of order N $\leq$ 3. For bay- and channel-type modes it is known that the accuracy of the channel model solutions is not very high. So, higher order models ought to be analysed to test the convergence properties. To a certain extent we have done this, see Stocker & Hutter (1986b).

(iv) The MWR was used by us in conjunction with a harmonic shape function set. It would be worthwhile to test further complete sets which make use of the peculiar features of the TW-behavior detected so far.

b) *On the physical side*

(i) How are TW's generated by the wind ? In a first attempt, one may only add the wind shear at the surface. The problem then consists of solving the inhomogeneous TW-equation. This will answer the question as to how long-periodic wind forces will be distributed among the different long-periodic modes. It is probable that gravity waves should not be ignored here as it is unknown how much of the input energy is taken over by the gravity and topographic waves, respectively.

More interesting, however is to know, how fast TW's react to wind impulses and wind cessation. It is likely that viscous effects cannot be ignored. What is the spin-up and spin-down time of the different TW-modes ?

(ii) Another question concerns the asymptotic stability of the linear TW-solutions. Do non-linearities cause some of these solutions to become unstable ? To our knwoledge only Ball's (1965) analysis of the steady elliptical modes is known.

(iii) The equations of barotropic-baroclinic coupling of TW's were studied in Chapter 2 within the context of linearized equations. Modifications of linearized internal waves by TW's can and should be analysed using those equations. Internal waves often have large amplitudes of the order of the epilimnion thickness. This suggests the extension of the investigations to non-linear processes.

(iv) Are bay-type modes favoured features ? In other words, in a lake consisting of several arms and bays (at the end of which topographic restoring is generally large), are bay-modes more easily excited than others ? Do the other modes, perhaps, have the faster spindown time ?

(v) For particle tracking it would be interesting to know how particle paths depend on the vorticity of the modes generated by wind input.

8.6 Measurements, observations

We close this brief list by a final remark regarding experiments and observations. In view of the dense population of the eigenfrequencies in a rather narrow frequency band it follows that detection of TW-motion by velocity or temperature time series from moored instruments is very difficult. A proper identification of individual channel-type modes is probably not possible. A dense distribution of instruments in a bay or at the long end of a channel or an elongated lake may, however, provide us with enough spatial resolution that a bay-mode could be identified observationally. Ball-modes are probably equally difficult to identify as are channel-modes.

Appendix A

In this Appendix we demonstrate how equations (2.66) can be reduced to a set of spatially two-dimensional equations by using a vertical shape function expansion of the velocity, pressure and density fields.

Consider the expansions

$$\tilde{u}(x,y,z,t) \;=\; \sum_{n=1}^{N} U_n(x,y,t)\; \phi_n\left(\tfrac{z}{H}\right),$$

$$\tilde{v}(x,y,z,t) \;=\; \sum_{n=1}^{N} V_n(x,y,t)\; \phi_n\left(\tfrac{z}{H}\right),$$

$$w(x,y,z,t) \;=\; \sum_{n=1}^{N} W_n(x,y,t)\; \Xi_n\left(\tfrac{z}{H}\right), \tag{A1}$$

$$p'(x,y,z,t) \;=\; \sum_{n=1}^{N} P_n(x,y,t)\; \psi_n\left(\tfrac{z}{H}\right),$$

$$\rho'(x,y,z,t) \;=\; \sum_{n=1}^{N} R_n(x,y,t)\; \chi_n\left(\tfrac{z}{H}\right),$$

Here $\{\phi_n,\ \Xi_n,\ \psi_n,\ \chi_n\}$, $n = 1,2,\ldots,N$ are a set of known functions of the independent variable z/H; through $H = H(x,y)$ they depend implicitly on x and y. It will be explained in the main text from which function set they will be chosen. The coefficient functions U_n, V_n, W_n, P_n, R_n depend on the spatial horizontal coordinates only and on the time. It is our goal to use (2.66) and the *Principle of Weighted Residuals* to deduce field equations for these quantities.

Let $\ll f_m,\ g_n \gg$ and $< f_m, g_n >$ be the following inner products:

$$\ll f_m,\ g_n \gg \;=\; \int_{-H}^{0} f_m\left(\tfrac{z}{H}\right)\, g_n\left(\tfrac{z}{H}\right)\, dz,$$

$$< f_m,\ g_n > \;=\; \int_{0}^{1} f_m(\xi)\, g_n(\xi)\, d\xi. \tag{A2}$$

They can be connected by the transformation

$$z = H(\xi - 1), \qquad dz = H\, d\xi. \tag{A3}$$

As a side remark we might mention that it would be more adequate to define $\ll,\gg$ as the integral from $z = -H$ to $z = \zeta$ and dropping terms involving ζ afterwards. This definition will be used in the transformation of (A12) below.

With the aid of (A2) and (A3) the reader may easily deduce the following properties:

$$<< f_m , g_n >> \ = \ H < f_m , g_n > ,$$

$$<< \frac{df_m}{dz} , g_n >> \ = \ < \frac{df_m}{d\xi} , g_n > ,$$

$$<< \frac{\partial f_m}{\partial x} , g_n >> \ = \ - \frac{\partial H}{\partial x} < \frac{df_m}{d\xi} (\xi-1) , g_n > ,$$

$$<< \frac{\partial f_m}{\partial y} , g_n >> \ = \ - \frac{\partial H}{\partial y} < \frac{df_m}{d\xi} (\xi-1) , g_n > .$$

(A4)

Consider now the momentum equations $(2.66)_{1,2}$ first. Substitute the expansions (A1) for $\tilde{u}$, $\tilde{v}$ and p' and form the following inner products: $<< (2.66)_1 , \delta\phi_m^M >>, << (2.66)_2 , \delta\phi_m^M >>$. We then obtain

$$A_{mn}^M \left[\frac{\partial U_n}{\partial t} - f V_n \right] - \frac{1}{\rho_*} \frac{\partial H/\partial x}{H} B_{mn} P_n$$

$$+ \frac{C_{mn}}{\rho_*} \frac{\partial P_n}{\partial x} = (g \frac{\partial \zeta}{\partial x} - \frac{\tau_1}{\rho_* H}) D_m^M ,$$

$$A_{mn} \left[\frac{\partial V_n}{\partial t} + f U_n \right] - \frac{1}{\rho_*} \frac{\partial H/\partial y}{H} B_{mn} P_n$$

$$+ \frac{C_{mn}}{\rho_*} \frac{\partial P_n}{\partial y} = (g \frac{\partial \zeta}{\partial y} - \frac{\tau_2}{\rho_* H}) D_m^M ,$$

$$(m = 1,2,3,\ldots,N),$$

(A5)

where

$$A_{mn}^M \ = \ < \delta\phi_m^M , \phi_n > ,$$

$$B_{mn} \ = \ < \delta\phi_m^M , (\xi-1) \frac{d\psi_n}{d\xi} > - < 1, (\xi-1)\frac{d\psi_n}{d\xi} > < 1, \delta\phi_m^M > ,$$

$$C_{mn} \ = \ < \delta\phi_m^M , \psi_n > - < 1, \psi_n > < 1, \delta\phi_m^M > ,$$

$$D_m^M \ = \ < 1, \delta\phi_m^M > .$$

(A6)

Here and henceforth, functions carrying a prefix δ are weighting functions which need not be the same as the shape functions in the expansions (A1). Moreover, in (A5) summation over repeated indices n is understood. In an analogous manner equations $(2.66)_{4,5}$ can be treated: The inner products $<< (2.66)_4 , \delta\phi_m^V >>$ and $<< (2.66)_5 , \delta\phi_m^V >>$ are formed and yield the equations

$$E_{mn} \frac{\partial R_n}{\partial t} - \frac{\rho_* N_{max}^2}{g} F_{mn} W_n = 0,$$

$$G_{mn} \frac{\partial W_n}{\partial t} + \frac{g}{\rho_*} E_{mn} R_n + \frac{1}{\rho_* H} H_{mn} P_n = 0,$$

$$(m = 1,2,3,\ldots,N),$$

(A7)

in which

$$E_{mn} = <\delta\phi_m^V, \chi_n>,$$

$$F_{mn} = <\delta\phi_m^V, \hat{N}^2 \Xi_n>,$$

$$G_{mn} = <\delta\phi_m^V, \Xi_n>, \tag{A8}$$

$$H_{mn} = <\delta\phi_m^V, \frac{d\psi_n}{d\xi}>,$$

and where

$$\hat{N}^2 = \hat{N}^2(\xi) = \frac{N^2(\xi)}{N_{max}^2} = -\frac{\dfrac{d\rho_0}{dz}}{\left|\dfrac{d\rho_0}{dz}\right|_{max}}. \tag{A9}$$

Clearly, in order that these relations are meaningful the boundary conditions $p'(\xi=1) = 0$ must be satisfied. Hence, we must request

$$\psi_n(\xi = 1) = 0. \tag{A10}$$

With our choice of ψ_n this condition will be nearly satisfied. Equations (A7) can be combined to yield the single equation

$$G_{mn} \frac{\partial^2 W_n}{\partial t^2} + N_{max}^2 F_{mn} W_n + \frac{H_{mn}}{\rho_* H} \frac{\partial P_n}{\partial t} = 0, \tag{A11}$$

$$(m = 1,2,3,\ldots,N).$$

Equations (A5) and (A11) are 3N equations for the unknowns U_n, V_n, P_n, W_n. The remaining N equations follow from the continuity equation $(2.66)_3$. Because kinematic boundary conditions at the upper and lower surfaces must be incorporated when employing the Principle of Weighted Residuals, we shall go into greater details. Forming $<< (2.66)_3, \delta\phi_m^C >>$ it follows that

$$<< \tilde{u}_x, \delta\phi_m^C >> + << \tilde{v}_y, \delta\phi_m^C >> + << w_z, \delta\phi_m^C >>$$
$$= (\frac{H_x}{H} \bar{u} + \frac{H_y}{H} \bar{v}) << 1, \delta\phi_m^C >>. \tag{A12}$$

Using the definition of the inner product $<< , >>$, we may easily prove that [*]

$$<< \tilde{u}_x, \delta\phi_m^C >> = \frac{\partial}{\partial x} << \tilde{u}, \delta\phi_m^C >> - << \tilde{u}, \frac{\partial \delta\phi_m^C}{\partial x} >>$$
$$- \tilde{u} \delta\phi_m^C \Big|_{z=\zeta} \frac{\partial \zeta}{\partial x} - \tilde{u} \delta\phi_m^C \Big|_{z=-H} \frac{\partial H}{\partial x},$$

$$<< \tilde{v}_y, \delta\phi_m^C >> = \frac{\partial}{\partial y} << \tilde{v}, \delta\phi_m^C >> - << \tilde{v}, \frac{\partial \delta\phi_m}{\partial y} >> \tag{A13}$$
$$- \tilde{v} \delta\phi_m^C \Big|_{z=\zeta} \frac{\partial \zeta}{\partial y} - \tilde{v} \delta\phi_m^C \Big|_{z=-H} \frac{\partial H}{\partial y},$$

$$<< \tilde{w}_z, \delta\phi_m^C >> = - << w, \delta\phi_{mz}^C >> + w \delta\phi_m^C \Big|_{z=\zeta} - w \delta\phi_m^C \Big|_{z=-H}.$$

---

[*] *Here we use the definition* $<< f, g >> = \int_{-H}^{\zeta} fg \, dz.$

With these expressions the weighted continuity statement (A11) takes on the form

$$\frac{\partial}{\partial x} \ll \tilde{u}, \delta\phi_m^C \gg + \frac{\partial}{\partial y} \ll \tilde{v}, \delta\phi_m^C \gg - \ll \tilde{u}, \frac{\partial\delta\phi_m^C}{\partial x} \gg - \ll \tilde{v}, \frac{\partial\delta\phi_m^C}{\partial y} \gg - \ll w, \delta\phi_{m_z}^C \gg$$

$$- \delta\phi_m^C(\zeta)\left(\left[u \zeta_x + v \zeta_y - w\right]_{z=\zeta} - \bar{u} \zeta_x - \bar{v} \zeta_y\right)$$

$$- \delta\phi_m^C(-H)\left(\left[u H_x + v Hy + w\right]_{z=-H} - \bar{u} H_x - \bar{v} H_y\right)$$

$$= \left(\frac{H_x}{H} \bar{u} + \frac{H_y}{H} \bar{v}\right) \ll 1, \delta\phi_m^C \gg .$$

(A14)

This equation is written down in full in order to demonstrate incorporation of the boundary conditions. The term in brackets in the second line equals $-\partial\zeta/\partial t$ and that in the third line vanishes. After this substitution we may ignore the two terms involving $\partial\zeta/\partial t$ because the rigid-lid assumption is made. Finally, the non-linear terms $\bar{u} \zeta_x$, $\bar{v} \zeta_y$ may be omitted because the non-linearities have consistently been dropped in earlier equations. Thus (A14) reduces to

$$\frac{\partial}{\partial x} \ll \tilde{u}, \delta\phi_m^C \gg + \frac{\partial}{\partial y} \ll \tilde{v}, \delta\phi_m^C \gg - \ll \tilde{u}, \frac{.\partial\delta\phi_m^C}{\partial x} \gg$$

$$- \ll \tilde{v}, \frac{\partial\delta\phi_m^C}{\partial y} \gg - \ll w, \delta\phi_{m_z}^C \gg$$

(A15)

$$= (\bar{u} H_x + \bar{v} H_y)\left[\frac{1}{H} \ll 1, \delta\phi_m^C \gg - \delta\phi_m^C(-H)\right].$$

Substituting the expansion (A1) and making use of formulas (A4) at appropriate places yields the equation

$$A_{mn}^C\left[\frac{\partial U_n}{\partial x} + \frac{\partial V_n}{\partial y}\right] - K_{mn}\left[\frac{H_x}{H} U_n + \frac{H_y}{H} V_m\right] - L_{mn} \frac{W_n}{H}$$

$$= (\bar{u} \frac{H_x}{H} + \bar{v} \frac{H_y}{H})\left[D_m^C - \delta\phi_m^C (\xi=0)\right], \qquad (m=1,2,3,\ldots,N),$$

(A16)

in which

$$A_{mn}^C = < \delta\phi_m^C, \phi_n >,$$

$$D_m^C = < \delta\phi_m^C, 1>,$$

$$K_{mn} = < (\xi-1)\frac{d\phi_n}{d\xi}, \delta\phi_m^C > - \delta\phi_m^C(0) \delta\phi_n^C(0),$$

$$L_{mn} = < \Xi_n, \frac{d\delta\phi_m^C}{d\xi} > .$$

(A17)

This completes the derivation of the baroclinic equations. They are: (A5), (A11) and (A16) and form 4N equations for the 4N baroclinic variables U_n, V_n, W_n, P_n, needless to say that the barotropic quantities $\bar{u}$, $\bar{v}$, ζ are regarded as being prescribed or governed by equations (2.61) or (2.70) in the main text.

Appendix B

Matrix elements $K^{ij\cdots}_{\beta\alpha}$ for $\varepsilon = 0.05$, $q = 2.0$

In this appendix we list the matrix elements $K^{ij\cdots}_{\beta\alpha}$ given by (5.3) up to the fourth order, thus $\beta, \alpha = 1, \ldots, 4$. They are calculated on a CYBER computer using the IMSL-library with a relative accuracy of 10^{-6}. Elements for further values of the bathymetric parameters are given in Stocker & Hutter (1985).

$$\varepsilon = 0.05 \qquad q = 2.0$$

$$K^{00++} = \begin{bmatrix} .57043E+00 & -.14127E+00 & .76439E-01 & -.49958E-01 \\ -.14127E+00 & .78816E+00 & -.26767E+00 & .16220E+00 \\ .76439E-01 & -.26767E+00 & .87392E+00 & -.33059E+00 \\ -.49958E-01 & .16220E+00 & -.33059E+00 & .92236E+00 \end{bmatrix}$$

$$K^{00--} = \begin{bmatrix} .71172E+00 & -.21771E+00 & .12640E+00 & -.85759E-01 \\ -.21771E+00 & .83812E+00 & -.30347E+00 & .18932E+00 \\ .12640E+00 & -.30347E+00 & .90104E+00 & -.35191E+00 \\ -.85759E-01 & .18932E+00 & -.35191E+00 & .93960E+00 \end{bmatrix}$$

$$K^{10++} = \begin{bmatrix} .46508E+00 & .74064E-01 & -.15149E+00 & .19250E+00 \\ -.90548E+00 & .12303E+01 & .41243E+00 & -.57411E+00 \\ .85049E+00 & -.25430E+01 & .17919E+01 & .89627E+00 \\ -.79269E+00 & .23963E+01 & -.40106E+01 & .22270E+01 \end{bmatrix}$$

$$K^{10--} = \begin{bmatrix} .88079E+00 & .21994E+00 & -.34679E+00 & .41375E+00 \\ -.17504E+01 & .15304E+01 & .64023E+00 & -.82680E+00 \\ .16460E+01 & -.32938E+01 & .20223E+01 & .11756E+01 \\ -.15369E+01 & .31098E+01 & -.46993E+01 & .24101E+01 \end{bmatrix}$$

$$K^{01++} = \begin{bmatrix} .46508E+00 & -.90548E+00 & .85049E+00 & -.79269E+00 \\ .74064E-01 & .12303E+01 & -.25430E+01 & .23963E+01 \\ -.15149E+00 & .41243E+00 & .17919E+01 & -.40106E+01 \\ .19250E+00 & -.57411E+00 & .89627E+00 & .22270E+01 \end{bmatrix}$$

$$K^{01--} = \begin{bmatrix} .88079E+00 & -.17504E+01 & .16460E+01 & -.15369E+01 \\ .21994E+00 & .15304E+01 & -.32938E+01 & .31098E+01 \\ -.34679E+00 & .64023E+00 & .20223E+01 & -.46993E+01 \\ .41375E+00 & -.82680E+00 & .11756E+01 & .24101E+01 \end{bmatrix}$$

$$K^{20+-} = \begin{bmatrix} -.23529E+01 & -.11459E+01 & .10217E+01 & -.10463E+01 \\ .19709E+01 & -.64457E+01 & -.26572E+01 & .22675E+01 \\ -.17666E+01 & .54644E+01 & -.10145E+02 & -.44194E+01 \\ .16171E+01 & -.49410E+01 & .86079E+01 & -.13603E+02 \end{bmatrix}$$

$$K^{20-+} = \begin{bmatrix} .52307E+00 & -.48178E+00 & .50338E+00 & -.52456E+00 \\ .44649E+01 & .18635E+01 & -.16195E+01 & .16279E+01 \\ -.37741E+01 & .83320E+01 & .35128E+01 & -.29585E+01 \\ .34021E+01 & -.70698E+01 & .11899E+02 & .53688E+01 \end{bmatrix}$$

$$K^{02+-} = \begin{bmatrix} .52307E+00 & .44649E+01 & -.37741E+01 & .34021E+01 \\ -.48178E+00 & .18635E+01 & .83320E+01 & -.70698E+01 \\ .50338E+00 & -.16195E+01 & .35128E+01 & .11899E+02 \\ -.52456E+00 & .16279E+01 & -.29585E+01 & .53688E+01 \end{bmatrix}$$

$$K^{02-+} = \begin{bmatrix} -.23529E+01 & .19709E+01 & -.17666E+01 & .16171E+01 \\ -.11459E+01 & -.64457E+01 & .54644E+01 & -.49410E+01 \\ .10217E+01 & -.26572E+01 & -.10145E+02 & .86079E+01 \\ -.10463E+01 & .22675E+01 & -.44194E+01 & -.13603E+02 \end{bmatrix}$$

$$K^{11++} = \begin{bmatrix} .28590E+01 & -.67465E+01 & .84240E+01 & -.95013E+01 \\ -.67465E+01 & .20655E+02 & -.28820E+02 & .30715E+02 \\ .84240E+01 & -.28820E+02 & .51820E+02 & -.61542E+02 \\ -.95013E+01 & .30715E+02 & -.61542E+02 & .95418E+02 \end{bmatrix}$$

$$K^{11--} = \begin{bmatrix} .99721E+01 & -.16406E+02 & .18664E+02 & -.20233E+02 \\ -.16406E+02 & .34649E+02 & -.43887E+02 & .44555E+02 \\ .18664E+02 & -.43887E+02 & .72092E+02 & -.81749E+02 \\ -.20233E+02 & .44555E+02 & -.81749E+02 & .12177E+03 \end{bmatrix}$$

$$K^{22++} = \begin{bmatrix} .15591E+02 & -.25701E+02 & .32091E+02 & -.36195E+02 \\ -.25701E+02 & .12098E+03 & -.10979E+03 & .11701E+03 \\ .32091E+02 & -.10979E+03 & .31491E+03 & -.23445E+03 \\ -.36195E+02 & .11701E+03 & -.23445E+03 & .59379E+03 \end{bmatrix}$$

$$K^{22--} = \begin{bmatrix} .56788E+02 & -.62500E+02 & .71102E+02 & -.77079E+02 \\ -.62500E+02 & .20719E+03 & -.16719E+03 & .16973E+03 \\ .71102E+02 & -.16719E+03 & .44383E+03 & -.31142E+03 \\ -.77079E+02 & .16973E+03 & -.31142E+03 & .76468E+03 \end{bmatrix}$$

$$K^{12+-} = \begin{bmatrix} -.11222E+02 & .30526E+02 & -.34717E+02 & .38281E+02 \\ .15136E+02 & -.57303E+02 & .91226E+02 & -.86801E+02 \\ -.17690E+02 & .58638E+02 & -.13211E+03 & .17984E+03 \\ .19376E+02 & -.61841E+02 & .11913E+03 & -.23436E+03 \end{bmatrix}$$

$$K^{12-+} = \begin{bmatrix} -.11222E+02 & .15136E+02 & -.17690E+02 & .19376E+02 \\ .30526E+02 & -.57303E+02 & .58638E+02 & -.61841E+02 \\ -.34717E+02 & .91226E+02 & -.13211E+03 & .11913E+03 \\ .38281E+02 & -.86801E+02 & .17984E+03 & -.23436E+03 \end{bmatrix}$$

We prove here that the dispersion curve $k(\sigma)$ for TW's in finite width channels is multi (double) valued if h'/h is bounded away from infinity everywhere within the channel. To this end we may start from the boundary value problem (5.11)

$$F" - \frac{h'}{h} F' - (k^2 - \frac{h'}{h} \frac{k}{\sigma}) f = 0, \qquad 0 < y < 1,$$

$$F = 0, \qquad y = 0,1,$$

(C1)

in which $h'/h < \infty$ and primes indicate differentiations with respect to y. For finite k and k/σ (C1) is a regular differential equation with analytic solutions for $y \in [0,1]$.

Solutions for k → 0: When k is small regular perturbation solutions may be sought as follows:

$$F = F_0 + k F_1 + k^2 F_2 + \ldots ,$$

$$\sigma = k \sigma_1^0 + k^2 \sigma_2^0 + \ldots , \qquad \text{as} \quad k \to 0.$$

(C2)

The zeroth order term in the expansion of σ is omitted because it would lead to an inconsistency. Substituting (C2) into (C1) yields the eigenvalue problem

$$F_0" - \frac{h'}{h} F_0' + \frac{h'}{h} \frac{1}{\sigma_1^0} F_0 = 0, \qquad 0 < y < 1,$$

$$F_0 = 0, \qquad y = 0,1$$

(C3)

for F_0 and a series of boundary value problems for F_n, $n \geq 1$, the first of which is given by

$$F_1" - \frac{h'}{h} F_1' + \frac{1}{\sigma_1^0} \frac{h'}{h} F_1 = \frac{h'}{h} \frac{\sigma_2^0}{\sigma_1^0} F_0, \qquad 0 < y < 1,$$

$$F_1 = 0, \qquad y = 0,1.$$

(C4)

Consider (C3) first. With the unique transformation

$$\xi = \int_0^y h \ (c\hat{y}) \ d\hat{y}, \qquad \xi \in [0, \xi_e = \int_0^1 h \ d\hat{y}]$$

(C5)

the eigenvalue problem can be transformed to Sturm-Liouville type,

$$\frac{d^2 F_0}{\partial \xi^2} + \frac{1}{\sigma_1^0} \frac{S(\xi)}{h(\xi)} \cdot F_0 = 0, \qquad 0 < \xi < \xi_e,$$

$$F_0 = 0, \quad \xi = 0, \xi_e,$$

(C6)

in which $h > 0$ and where

$$S(\xi) = \frac{1}{h(\xi)} \frac{dh}{d\xi}$$

(C7)

is the transformed slope parameter. It follows that there is a countable set of real eigenvalues $\sigma_1^{0\,(m)}$, $m = 1,2,3,\ldots$, with associated eigenfunctions $F_0^{(m)}$ which form a complete function set. Moreover, from the Rayleight quotient for the mth mode

$$(\sigma_1^0)^{(m)} = \int_0^{\xi_e} \frac{S(\xi)}{h(\xi)} F_0^{(m)\,2}(\xi)\, ds / \int_0^{\xi_e} \left(\frac{d\,F_0^{(m)}}{d\xi}\right) d\xi, \qquad m = 1,2,3,\ldots$$

we may deduce that $\sigma_1 > 0$ whenever $S > 0$ for all $\xi \in [0, \xi_e]$. Since qualitatively for $k/\sigma > 0$ F_0 is oscillatory where $S > 0$ and exponentially decaying where $S < 0$, the eigenvalues $(\sigma_1^0)^{(m)}$ are, however, positive for a much larger class of slope factors S. It also follows from (C6) that the eigenfunctions $F_0^{(m)}$ corresponding to different eigenvalues satisfy the orthogonality relations

$$\int_0^{\xi_e} \frac{S(\xi)}{h(\xi)} F_0^{(m)}(\xi)\, F_0^{(n)}(\xi)\, d\xi = \begin{cases} 0, & m \neq n, \\ 1, & m = n. \end{cases} \qquad (C8)$$

These considerations prove that, asymptotically, $\sigma \sim \sigma_1^0 k$ as $k \to 0$.

To obtain the next higher order term, invoke (C5) in (C4) to obtain the TPBVP

$$\frac{d^2 F_1}{d\xi^2} + \frac{1}{\sigma_1^{0\,(m)}} \frac{S}{h} F_1 = \frac{\sigma_2^0}{\sigma_1^0} \frac{S}{h} F_0^{(m)}, \qquad 0 < \xi < \xi_e,$$

$$F_1(0) = F_1(\xi_e) = 0. \qquad (C9)$$

The left hand side of this equation is identical to $(C6)_1$. So, the inhomogeneous problem can only possess a solution if the right hand side satisfies the orthogonality condition

$$\frac{\sigma_2^0}{\sigma_1^{0\,(m)}} \int_0^{\xi_e} \frac{S}{h} (F_0^{(m)}(\xi))^2\, d\xi = 0,$$

which according to (C8) is only possible if $\sigma_2^0 = 0$. Hence

$$\sigma \sim \sigma_1^0 k + O(k^3), \qquad \text{as} \quad k \to 0. \qquad (C10)$$

Further terms will not be determined.

There is a second, perhaps more direct approach to construct asymptotic solutions from (C1). To this end we multiply equation (C1) by F and subsequently integrate the resulting equation from $y = 0$ to $y = 1$. With integrations by parts used in appropriate terms this yields

$$A k^2 + \frac{k}{\sigma} B + C = 0 \qquad (C11)$$

with
$$A = \int_0^1 h^{-1}\, F^2\, dy, \qquad B = \int_0^1 (h^{-1})'\, F^2\, dy, \qquad C = \int_0^1 h^{-1}\, (F')^2\, dy.$$

The integrals A, B, C are non-zero and bounded as long as h·and h' are bounded, since F is analytic in this case. Moreover, for realistic topographies A and C have the same sign but the sign of B is opposite to that of A and B because F is right-bound. Moreover, because F depends on k, so do A, B, C, so that we may write

$$A = A_0^0 + A_1^0 k + \ldots , \text{ etc.,} \left.\begin{array}{c} \\ \\ \end{array}\right\} \text{ as } k \to 0. \qquad (C12)$$
$$\sigma = \sigma_0^0 + \sigma_1^0 k + \ldots ,$$

Substituting (C12) into (C11) and collecting terms of equal power in k shows that necessarily

$$\sigma_0^0 = 0, \qquad \sigma_1^0 = -\frac{B_0^0}{C_0^0}. \qquad (C13)$$

Hence

$$\sigma \sim \sigma_1^0 k \qquad \text{as} \qquad k \to 0, \qquad (C14)$$

as before. Substituting this result into (C11) yields the eigenvalue problem

$$(h^{-1}F')' + \frac{1}{\sigma_1^0}(h^{-1})' F = 0, \qquad 0 < y < 1, \left.\begin{array}{c} \\ \\ \\ \end{array}\right\} \text{ as } k \to 0$$
$$F = 0, \quad y = 0,1,$$

correct to O(k). This eigenvalue problem determines σ_1^0 and ·corresponds to (C6).

Solutions for $k \to \infty$: A detailed proof for arbitrary bounded h'(y) and h(y) > 0 is complicated, but when A, B have signs opposite to that of C the behavior of σ as k → ∞ can easily be obtained from equation (C11) by substituting the asymptotic series

$$A = A_0^\infty + A_1^\infty \frac{1}{k} + \ldots , \text{ etc.,} \left.\begin{array}{c} \\ \\ \end{array}\right\} \text{ as } k \to \infty. \qquad (C15)$$
$$\sigma = \sigma_0^\infty + \sigma_1^\infty \frac{1}{k} + \ldots ,$$

We then obtain from (C11)

$$\sigma_0^\infty = 0, \qquad \sigma_1^\infty = -\frac{B_0^\infty}{A_0^\infty} \qquad (C16)$$

and hence deduce

$$\sigma \sim -\frac{B_0^\infty}{A_0^\infty}\frac{1}{k}, \qquad \text{as} \quad k \to \infty, \qquad (C17)$$

correct to O(1/k). Because σ depends continuously on k and is differentiable for $k \in [0,\infty)$, (C14) and (C17) imply that the frequency σ approaches zero for both, small and large wavenumbers. Consequently, there exists a wavenumber k_0 with zero group velocity.

There remains the determination of the eigenvalue σ_1^∞. Substitution of (C.17) into (C1) yields

$$F'' - \frac{h'}{h} F' - k^2 \left(1 + \frac{A_0^\infty}{B_0^\infty} \frac{h'}{h}\right) F = 0, \qquad 0 < y < 1,$$

$$F = 0, \qquad\qquad\qquad\qquad y = 0,1, \qquad\qquad \text{as } k \to \infty. \qquad (C18)$$

This eigenvalue problem for the eigenvalue A_0^∞/B_0^∞ must be solved by WKB-methods (see Bender & Orszag, 1978, Chapter 10). With the substitution $F(y) = h^{1/2}(y) \, G(y)$ it assumes its standard form.

A more systematical approach for the determination of σ when $k \to \infty$ is as follows. Substituting the transformation

$$F(y) = h^{1/2}(y) \, G(y) \qquad\qquad (C19)$$

into (C1) produces the new boundary value problem

$$G'' - \left[\left(k^2 - \frac{h'}{h} \frac{k}{\sigma}\right) + \frac{3}{4} \frac{h'^2}{h^2} + \frac{1}{2} \frac{h''}{h}\right] G = 0, \qquad 0 < y < 1,$$

$$B = 0, \quad y = 0,1. \qquad\qquad\qquad\qquad\qquad\qquad\qquad (C20)$$

If $\sigma = O(1)$ or larger no solution fulfilling the two boundary conditions can be found when $k \to \infty$. Thus we use the expansion

$$\sigma = \sigma_1^\infty \frac{1}{k} + \sigma_2^\infty \frac{1}{k^2} \, \cdots.$$

and then obtain from (C20)

$$G'' - \left[\chi_0(y) \, k^2 + \chi_1(y) \, k + O(1)\right] G = 0, \qquad 0 < y < 1,$$

$$G = 0, \qquad\qquad\qquad\qquad\qquad\qquad y = 0,1, \qquad \text{as } k \to \infty, \qquad (C21)$$

where

$$\chi_0(y) \equiv 1 - S(y) \frac{1}{\sigma_1^\infty},$$

$$\chi_1(y) \equiv \frac{\sigma_2^\infty}{(\sigma_1^\infty)^2} \, S(y). \qquad\qquad (C22)$$

In the above equations it was assumed that $S \ll k$ and $h''/h \ll k$. The dominant balance in (C22) corresponds to the equation

$$G'' - \chi_0(y) \, k^2 \, G = 0, \qquad 0 < y < 1,$$

$$G = 0, \qquad\qquad\qquad y = 0,1, \qquad \text{as } k \to \infty. \qquad (C23)$$

This equation is now in the appropriate form to apply the WKB-method, but detailed procedures depend on the number of turning points which $\chi_0(y)$ possesses. These turning points are given by the solution of the equation

$$\chi_0(y_T) = 0, \qquad \text{or} \qquad \sigma_1^\infty = S(y_T).$$

In regions, where $\chi_0 < 0$ the solution to (C23) is oscillatory; in regions where $\chi_0 > 0$ solutions are exponential. These solutions are connected at the turning points by transition solutions consisting of the Airy-function and the eigenvalue is determined through satisfaction of the boundary conditions. The details of this computation are very cumbersome even when there is only one single turning point.

For the determination of the main result, namely the behavior of $\sigma(k)$ as $k \to 0$ and $k \to \infty$, it suffices however to establish the relations (C14) and (C17). In view of the definitions of A, B and C and the analyticity of $F(y)$ we know that σ_1 and σ_1^∞ are finite nonzero numbers as long as h' and h are finite and h is bounded away from zero. The exact eigenvalues need not necessarily be constructed.

References

Abramowitz M.,   1972. *Handbook of mathematical functions.* Dover.
Stegun I.A.

Allen J.S.   1975. Coastal trapped waves in a stratified ocean.
J. Phys. Oceanogr., 5, 300-325.

Ball F.K.   1965. Second class motions of a shallow liquid. *J. Fluid Mech.*, 23, 545-561.

Ball F.K.   1967. Edge waves in an ocean of finite depth. *Deep-Sea Res.*, 14, 79-88.

Bäuerle E.   1984. Topographic waves in the Baltic Sea. *Proc. XIV. Conf. of Baltic Oceanographers, Gdynia.*

Bäuerle E.   1985. Internal free oscillations in the Lake of Geneva. *Annales Geophysicae,* 3, 199-206.

Bäuerle E.   1986. Eine Untersuchung über topographische Wellen in einem Kanalmodell. *Mitteilungen der Versuchsanstalt für Wasserbau, Hydrologie und Glaziologie, ETH-Zürich,* No. 83.

Bäuerle E.,   1986. A numerical study of topographic waves in a chan-
Hutter K.   nel on the f-plane. *J. Phys. Oceanogr.,* (submitted).

Bender C.M.,   1978. *Advanced Mathematical Methods for Scientists and Engineers.*
Orszag S.A.   McGraw Hill.

Bennett J.R.,   1981. Calculation of the rotational normal modes of
Schwab D.J.   oceans and lakes with general orthogonal coordinates. *J. Comput. Phys.,* 44, 359-376.

Brink K.H.   1980. Propagation of barotropic continental shelf waves over irregular bottom topography. *J. Phys. Oceanogr.,* 10, 765-778.

Brink K.H.   1982. A comparison of long coastal trapped wave theory with observations off Peru. *J. Phys. Oceanogr.,* 12, 897-913.

Brown P.J.   1973. Kelvin wave reflection in a semi-infinite canal. *J. Mar. Res.,* 31, 1-10.

Buchwald V.T., 1968. The propagation of continental shelf waves. *Proc.*
Adams J.K.   *Roy. Soc.,* A305, 235-250.

Buchwald V.T., 1977. Resonance of shelf waves near islands, in *Lecture*
Melville W.K. *Notes in Physics,* vol. 64, edited by D.G. Provis and R. Radok, 202-205, Springer, New York.

Buchwald V.T., 1975. Rectangular resonators on infinite and semi-infi-
Williams N.V. nite channels. *J. Fluid Mech.,* 67, 497-511.

Clarke A.J.   1977. Observational and numerical evidence for wind-forced coastal trapped long waves. *J. Phys. Oceanogr.,* 7, 231-247.

Courant R., Hilbert D.
1967. *Methoden der mathematischen Physik I, II.* Springer Verlag.

Csanady G.T.
1976. Topographic waves in Lake Ontario. *J. Phys. Oceanogr.*, 6, 93-103.

Djurfeldt L.
1984. A unified derivation of divergent second-class topographic waves. *Tellus*, 36 A, 306-312.

Duillier F. de
1730. Remarques sur l'histoire du lac de Genève in Spon. *Histoire de Genève*, 2, 463.

Ertel H.
1942. Ein neuer hydrodynamischer Wirbelsatz. *Meteorolog. Z.*, 59, 277-281.

Finlayson B.A.
1972. *The Method of Weighted Residuals and Variational Principles.* Academic Press.

Gill A.E., Schumann E.H.
1974. The generation of long shelf waves by the wind. *J. Phys. Oceanogr.*, 4, 83-90.

Graf W.
1983. Hydrodynamics of the Lake of Geneva. *Schw. Z. Hydrol.*, 45, 62-100.

Gratton Y.
1983. *Low frequency vorticity waves over strong topography.* Ph.D. thesis, Univ. of British Columbia, 132 pp.

Gratton Y., LeBlond P.H.
1986. Vorticity waves over strong topography. *J. Phys. Oceanogr.*, 16, 151-166.

Hamblin P.F.
1972. *Some free oscillations of a rotating natural basin.* Ph.D thesis, Univ. Washington, Dept. Oceanography, 97 pp.

Hogg H.G.
1980. Observations of internal Kelvin waves trapped round Bermuda. *J. Phys. Oceanogr.*, 10, 1353-1376.

Holton J.R.
1979. *An Introduction to Dynamical Meteorology*, 2nd ed., Academic Press.

Horn W., Mortimer C.H., Schwab D.J.
1986. Internal dynamics of the Lake of Zurich. *Limnology and Oceanography*, (in press).

Huang J.C.K., Saylor J.H.
1982. Vorticity waves in a shallow basin. *Dyn. Atmos. Oceans.*, 6, 177-196.

Huthnance J.M.
1975. On trapped waves over a continental shelf. *J. Fluid Mech.*, 69, 689-704.

Huthnance J.M.
1978. On coastal trapped waves: Analysis and numerical calculation by inverse iteration. *J. Phys. Oceanogr.*, 8, 74-92.

Hutter K.
1983. Strömungsdynamische Untersuchungen im Zürich- und Luganersee. *Schw. Z. Hydrol.*, 45, 102-144.

Hutter K.
1984a. Fundamental equations and approximations, in *Hydrodynamics of Lakes*, CISM 286, edited by K. Hutter, 1-37, Springer, Wien-New York.

Hutter K.   1984b. Linear gravity waves, in *Hydrodynamics of Lakes*,
                       CISM 286, edited by K. Hutter, 39-80, Springer, Wien-
                       New York.

Hutter K.   1984c. Mathematische Vorhersage von barotropen und ba-
                       roklinen Prozessen im Zürich- und Luganersee. *Viertel-
                       jahresschrift der Naturf. Ges.*, 129, 51-92.

Hutter K.,   1983. On internal wave dynamics in the northern basin
Salvadè G.,   of Lake of Lugano. *Geophys. Astrophys. Fluid Dyn.*, 27, 299-
Schwab D.J.   336.

Hutter K.,   1986. Lake Hydraulics, in *Developments in Hydraulic Engi-
Vischer D.   neering*, vol. 4, edited by P. Novak, Elsevier Appl.
                       Science Publ., Amsterdam (in press).

Johnson E.R.   1986. Topographic waves in elliptical basins. *Geophys.
                       Astrophys. Fluid Dyn.* (in press).

Kielmann J.   1983. The generation of eddy-like structures in a model
                       of the Baltic Sea by low frequency wind forcing (un-
                       published).

Kielmann J.,   1984. Baroclinic circulation models, in *Hydrodynamics of
Simons T.J.   Lakes*, CISM 286, edited by K. Hutter, 235-285, Springer,
                       Wien-New York.

Koutitonsky V.G.   1985. Subinertial coastal-trapped waves in channels
                       with variable stratification and topography. Ph.D.
                       thesis, Marine Sciences Res. Center, State Univ. New-
                       York, Stony Brook.

Lamb H.   1932. *Hydrodynamics*. 6th ed. Cambridge University Press.

Larsen J.C.   1969. Long waves along a single-step topography in a
                       semi-infinite uniformly rotating ocean. *J. Mar. Res.*,
                       27, 1-6.

LeBlond P.H.,   1980. *Waves in the Ocean*. Elsevier.
Mysak L.A.

Lie H.-J.   1983. Shelf waves on the exponential, linear and sinus-
                       oidal bottom topographies. *Bull: KORDI*, 5, 1-8.

Lie H.-J.,   1983. Formation of eddies and transverse currents in a
El-Sabh M.I.   two-layer channel of variable bottom: Applications to
                       the lower St. Lawrence estuary. *J. Phys. Oceanogr.*, 10,
                       1063-1075.

Marmorino G.O.   1979. Lowfrequency current fluctuation in Lake Ontario.
                       J. Geophys. Res., 84, 1206-1214.

Miles J.W.   1974. Harbor seiching. *Ann. Rev. Fluid Mech.*, 6, 17-35.

Miles J.W.,   1963. On free-surface oscillations in a rotating para-
Ball F.K.   boloid. *J. Fluid Mech.*, 17, 257-266.

Miles J.W.,   1975. Helmholtz resonance of harbours. *J. Fluid Mech.*, 67,
Lee Y.K.   445-464.

Mortimer C.H., 1982. Internal wave dynamics and their implications for
Horn W.   plankton biology in the Lake of Zurich. *Vierteljahresschr.*
                 der Naturf. Ges. Zürich, 127, 299-318.

Mysak L.A.   1967. On the theory of continental shelf waves. *J. Mar.*
                 Res., 25, 205-227.

Mysak L.A.   1968. Edgewaves on a gently sloping continental shelf
                 of finite width. *J. Marine Res.*, 26, 24-33.

Mysak L.A.   1980a. Recent advances in shelf wave dynamics. *Reviews of*
                 Geophysics and Space Physics, 18, 211-241.

Mysak L.A.   1980b. Topographically trapped waves. *Ann. Rev. Fluid Mech.*,
                 12, 45-76.

Mysak L.A.   1984. Topographic waves in lakes, in *Hydrodynamic of Lakes*,
                 CISM 286, edited by K. Hutter, 81-128, Springer, Wien-
                 New York.

Mysak L.A.   1985. Elliptical topographic waves. *Geophys. Astrophys.*
                 Fluid Dyn., 31, 93-135.

Mysak L.A,,   1979. Trench waves. *J. Phys. Oceanogr.*, 9, 1001-1013.
LeBlond P.H.,
Emery W.J.

Mysak L.A.,   1983. Lake of Lugano and topographic waves. *Nature*, 306,
Salvadè G.,   46-48.
Hutter K.,
Scheiwiller T.

Mysak L.A.,   1985. Topographic waves in an elliptical basin, with
Salvadè G.,   application to the Lake of Lugano. *Phil. Trans. R. Soc.*
Hutter K.,   *London*, A 316, 1-55.
Scheiwiller T.

Ou H.W.   1980. On the propagation of the topographic Rossby
                 waves near continental margins. Part 1: Analytical mo-
                 del for a wedge. *J. Phys. Oceanogr.*, 10, 1051-1060.

Pearson C.E.   1974. *Handbook of Applied Mathematics*. VNR Company.

Pedlosky J.   1982. *Geophysical Fluid Dynamics*. Springer Verlag, New York.

Poincaré H.   1910. Théorie des marées. *Leçons de mécanique céleste*, 3,
                 Paris.

Protter M.H.,   1967. *Maximum Principles of Differential Equations*. Prentice-
Weinberger H.F. Hall Inc. New Jersey.

Raggio G.,   1982. An extended channel model for the prediction of
Hutter K.   motion in elongated homogeneous lakes. *J. Fluid Mech.*,
                 121, 231-299.

Rao D.B.,   1976. Two-dimensional normal modes in arbitrary enclo-
Schwab D.J.   sed basins on a rotating earth: Applications to Lakes
                 Ontario and Superior. *Phil. Trans. R. Soc.*, A281, 63-96.
                 63-96.

Reid R.O.   1958. Effect of Coriolis force on edge waves. 1: Inve-
                     stigation of the normal modes. *J. Mar. Res.*, 16, 104-144.

Rhines P.B.   1969. Slow oscillations in an ocean of varying depth.
                     J. Fluid Mech., 37, 161-205.

Ripa P.   1978. Normal modes of a closed basin with topography,
                     J. Geophys. Res., 83, 1947-1957.

Saint-Guily B.   1972. Oscillations propres dans un bassin de profon-
                     deur variable: Modes de seconde classe, in *Studi in Onore
                     di Giuseppina Aliverti*, 15-25. Instituto Universitario
                     Navale di Napoli, Napoli, Italy.

Saylor J.H.,   1980. Vortex modes in Southern Lake Michigan. *J. Phys.
Huang J.S.K.,   Oceanogr.*, 10, 1814-1823.
Reid R.O.

Saylor J.H.,   1983. *Vortex Modes of Particular Great Lakes Basins*. Great
Miller G.S.   Lakes Environmental Research Laboratory, Contribution
                     No. 394, NOAA, Ann Arbor, Michigan.

Scheiwiller T.,   1986. Dynamics of powder snow avalanches. *J. Fluid Mech.*
Hutter K.,   (submitted).
Hermann F.

Sezawa K.,   1939. On shallow water waves transmitted in the direc-
Kanai K.   tion parallel to a sea coast, with special reference
                     to long-waves in heterogeneous media. *Bull. Earthquake
                     Res. Inst. Tokyo*, 17, 685-694.

Simons T.J.   1978a. Wind-driven circulations in the south-west Bal-
                     tic. *Tellus*, 30, 272-283.

Simons T.J.   1978b. Generation and propagation of downwelling
                     fronts. *J. Phys. Oceanogr.*, 8, 571-581.

Simons T.J.   1980. Circulation models of lakes and inland seas. *Can.
                     Bull. Fish. Aquat.*, Sci., 203, 1-146.

Simons T.J.   1983. Resonant topographic response of nearshore cur-
                     rents to wind forcing. *J. Phys. Oceanogr.*, 13, 512-523.

Snodgrass F.E.,   1962. Long-period waves over California's continental
Munk W.H.,   borderland. I: Background spectra. *J. Mar. Res.*, 20,
Miller G.R.   3-30.

Stocker T.,   1985. A model for topographic Rossby waves in channels
Hutter K.   and lakes. *Mitteilungen der Versuchsanstalt für Wasserbau,
                     Hydrologie und Glaziologie*, ETH Zürich, No. 76.

Stocker T.,   1986a. One-dimensional models for topographic Rossby
Hutter K.   waves in elongated basins on the f-plane. *J. Fluid Mech.*,
                     170, 435-459.

Stocker T.,   1986b. Topographic Rossby waves in rectangular basins.
Hutter K.   *J. Fluid Mech.*, (submitted).

Szidarovszky F., Yakowitz S.	1978. *Principles and Procedures of Numerical Analysis*. Plenum Press, New York.
Takeda H.	1984. Topographically trapped waves over the continental shelf and slope. *J. Oceanogr. Soc. Japan*, 40, 349-366.
Taylor G.I.	1922. Tidal oscillations in gulfs and rectangular basins. *Proc. London Math. Soc.*, Ser. 2, 148-181.
Trösch J.	1984. Finite element calculation of topographic waves in lakes. *Proceedings of 4th International Conference on Applied Numerical Modeling, Tainan, Taiwan.*
Wang D.P., Mooers C.N.K.	1976. Coastal trapped waves in a continuously stratified ocean. *J. Phys. Oceanogr.*, 6, 853-863.
Webb A.J., Pond S.	1986. The propagation of a Kelvin wave around a bend in a channel. *J. Fluid Mech.*, 169, 257-274.
Wenzel M.	1978. *Interpretation der Wirbel im Bornholmbecken durch topographische Rossby Wellen in einem Kreisbecken.* Diplomarbeit, Christian Albrechts Universität Kiel, 52 pp.

Author index

Subject index

The following abbreviations are used: MWR = Method of Weighted Residuals,
TW = topographic wave.